BANK FEES:
CREEPING UP AND UP

BANK FEES: CREEPING UP AND UP

U.S. GOVERNMENT
ACCOUNTABILITY OFFICE

Novinka Books
New York

For permission to use material from this book please contact us:
Telephone 631-231-7269; Fax 631-231-8175
Web Site: http://www.novapublishers.com

NOTICE TO THE READER

LIBRARY OF CONGRESS CATALOGING-IN-PUBLICATION DATA
United States. Government Accountability Office.
Bank fees : creeping up & up / GAO. p. cm.
ISBN 978-1-60692-039-8 (softcover)
1. Banks and banking--Service charges--United States. I. Title.
HG1616.S4U55 2008
332.1'7--dc22 2008032040

Published by Nova Science Publishers, Inc. ≃ New York

CONTENTS

PREFACE

In 2006, consumers paid over $36 billion in fees associated with checking and savings accounts, raising questions about consumers' awareness of their accounts' terms and conditions. The author was asked to review (1) trends in the types and amounts of checking and deposit account fees since 2000, (2) how federal banking regulators address such fees in their oversight of depository institutions, and (3) the extent that consumers are able to obtain account terms and conditions and disclosures of fees upon request prior to opening an account. The author analyzed fee data from private data vendors, publicly available financial data, and information from federal regulators; reviewed federal laws and regulations; and used direct observation techniques at depository institutions nationwide. This is an excerpted and indexed version.

ABBREVIATIONS

ACH	Automated Clearing House
ATM	automated teller machine
CAESAR	Complaint Analysis Evaluation System and Reports
CCS	Consumer Complaint System
EFT	electronic funds transfer
FDIC	Federal Deposit Insurance Corporation
NCUA	National Credit Union Administration
OCC	Office of the Comptroller of the Currency
OTS	Office of Thrift Supervision
PIN	personal identification number
PIRG	U.S. Public Interest Research Group

SCDA	service charges on deposit accounts
STARS	Specialized Tracking and Reporting System
TFR	Thrift Financial Reports
TISA	Truth in Savings Act

BANK FEES

FEDERAL BANKING REGULATORS COULD BETTER ENSURE THAT CONSUMERS HAVE REQUIRED DISCLOSURE DOCUMENTS PRIOR TO OPENING CHECKING OR SAVINGS ACCOUNTS *

January 31, 2008
The Honorable Carolyn B. Maloney
Chairwoman
Subcommittee on Financial Institutions
and Consumer Credit
Committee on Financial Services
House of Representatives

Dear Chairwoman Maloney:

In 2006, consumers paid over $36 billion in various fees associated with checking and savings accounts at depository institutions—banks, thrifts, and credit unions.[1] Members of Congress, consumer groups, and others have raised a variety of concerns about these fees—for example, whether depository institutions have increased fees as a source of revenues and if so, the impact of this trend on consumers. Additionally, some have questioned how regulators address fee practices in their oversight of depository institutions and whether consumers, prior to opening a checking or savings

* This Book is an excerpted, indexed version of GAO Report GAO 08-281, Dated January 2008

account, are able to obtain information on fees and depository institution practices that influence when fees are assessed.

The Board of Governors of the Federal Reserve System (Federal Reserve) has established regulations for checking and savings accounts that require depository institutions to disclose certain information about the fees they charge. Specifically, Regulation DD, which implements the Truth in Savings Act (TISA), requires depository institutions to disclose (among other things) the amount of any fee that may be imposed in connection with an account and the conditions under which such fees are imposed.[2] Regulation E—the other primary federal regulation governing checking and savings account fees—implements the Electronic Fund Transfer Act and establishes the basic rights, liabilities, and responsibilities of consumers who use electronic fund transfer services and of financial institutions that offer these services.[3] To ensure compliance with these and other relevant laws and regulations, banks, thrifts, and credit unions are subject to oversight at the federal and state level.[4] This oversight includes on-site examinations and other steps to ensure compliance with the laws and regulations. In 2005, partly in response to concerns about the marketing, implementation, and fees of overdraft protection programs being offered by depository institutions, the OCC, Federal Reserve, FDIC and NCUA jointly and the OTS separately issued guidance (interagency guidance) outlining "best practices" that address, among other things, communicating the features of these programs to customers.[5]

You requested that we examine a number of issues related to the fees that consumers pay on their checking and savings accounts. This report discusses (1) the trends in the types and amounts of fees associated with checking and deposit accounts since 2000 and available information on the characteristics of consumers that incur fees; (2) ways that federal and selected state banking regulators address checking and deposit account fees in their oversight of depository institutions; and (3) the extent to which consumers are able to obtain information on account terms and conditions and on fees, including information about specific transactions and bank practices that determine when such fees are assessed, upon request prior to opening an account. In addition, Appendix II of the report presents information on issues related to providing real-time account information at point-of-sale terminals and automated teller machines (ATM) that could help consumers avoid certain fees.

For the first objective, we engaged the services of a private sector firm—Moebs $ervices, Inc.—to obtain data on selected fees associated with checking and savings accounts from 2000 to 2007 and similar data from another private sector firm—Informa Research Services, Inc.—from 2000 to 2006. We interviewed representatives of these two firms to understand their methodology for collecting the data and ensuring its integrity. In addition, we conducted reasonableness checks on the data we received to identify any missing, erroneous, or outlying data and concluded that the data were sufficiently reliable for use in our report. To determine the role that these fees have played in depository institutions' revenues, we also obtained and analyzed quarterly financial data submitted by federally insured banks, thrifts, and credit unions and maintained by FDIC and NCUA. In our past work, we have found the quarterly financial data maintained by FDIC and NCUA to be sufficiently reliable for the purposes of our reports. We also reviewed the literature for studies or information on the characteristics of consumers who might be likely to incur such fees and interviewed representatives of the federal banking regulators about this issue. To determine how federal and selected state banking regulators address fees associated with checking and deposit accounts as part of their oversight of depository institutions, we obtained and reviewed examination manuals and guidance used by the five federal banking regulators and state regulators in six states.[6] We obtained and reviewed a sample of 25 reports on examinations conducted during 2006 to identify how these regulators carried out examinations for compliance with Regulations DD and E.[7] In addition, we obtained data from each of the federal banking regulators on violations they cited for institutions' non-compliance with Regulation DD and Regulation E disclosure-related provisions, as well as enforcement actions that each regulator took against institutions from 2002 to 2006. We also obtained annual data on consumer complaints concerning checking and savings accounts at depository institutions—particularly complaints related to fees and disclosures—as well as complaints for other major products (credit cards and mortgage loans) referred to these regulators from 2002 to 2006. To assess the reliability of data from the five federal banking regulators, we reviewed relevant documentation and interviewed agency officials. Finally, we interviewed officials from each of the federal banking regulators and from six state banking regulators about these issues.

To assess the extent to which consumers are able to obtain account terms and conditions and disclosures of fees, we used direct observation techniques and reviewed studies and reports by government agencies, consumer groups, and other researchers. We also reviewed relevant federal laws, regulations, and guidance issued by the federal banking regulators. For direct observation, GAO employees posed as consumers shopping for checking and savings accounts and visited 185 branches of 154 banks, thrifts, and credit unions throughout the nation to request documents on the fees associated with basic checking and savings accounts.[8] We selected these institutions to ensure a mix of institution type (bank, thrift, and credit union) and size; however, the results cannot be generalized to all institutions. These employees also reviewed information from the institutions' Web sites. To obtain information on issues related to providing consumers with real-time account information during debit card transactions at point-of-sale terminals and automated teller machines, we reviewed available literature from the Federal Reserve and other sources and met with officials from depository institutions, card associations, third-party processors, and trade organizations.

We conducted this performance audit from January 2007 to January 2008, in accordance with generally accepted government auditing standards. Those standards require that we plan and perform the audit to obtain sufficient, appropriate evidence to provide a reasonable basis for our findings and conclusions based on our audit objectives. We believe that the evidence obtained provides a reasonable basis for our findings and conclusions based on our audit objectives. Appendix I explains our objectives, scope, and methodology in greater detail.

RESULTS IN BRIEF

According to data from private vendors, average fees for some checking and savings account features—such as overdrafts, insufficient funds (instances in which an institution denies a transaction that would result in an overdraft but charges a fee), returns of deposited items, and stop payment orders—have generally risen since 2000, while others—for example, monthly account maintenance fees—have generally declined. For example, the average overdraft fee increased by about 11 percent (after inflation adjustment) between 2000 and 2007 among institutions surveyed by Moebs $ervices. The data also indicate some variation in fees by type and size of institution, with banks and thrifts charging higher fees on average than credit unions, and larger institutions charging more on average than midsize and smaller institutions. During this same period, the portion of income that depository institutions derived from non-interest sources—including, but not limited to, fees on savings and checking accounts—varied, but generally increased from about 24 percent to 27 percent of income from all sources. Changes in both consumer behavior and the practices of depository institutions are likely influencing these trends in fees. For example, consumers are increasingly using electronic forms of payment that result in rapid or even immediate debits—a development that may mean an increasing number of charges for insufficient funds or overdrafts. Additionally, many depository institutions have automated overdraft protection programs that have been increasingly marketed to customers. However, we were not able to analyze the demographic characteristics of customers that incur bank fees because doing so would require transaction-level data for all account holders—data that are not publicly available. FDIC is currently reviewing the overdraft programs of some of the banks it supervises, including reviewing transaction-level data to help determine the characteristics of consumers who incur fees related to overdrafts, but its study will not be completed until late 2008.

Federal banking regulators address fees associated with checking and savings accounts primarily by examining depository institutions' compliance with statutory and regulatory disclosure requirements and reviewing customer complaints. However, regulators generally do not address the reasonableness of fees assessed. The examination procedures for financial institutions' compliance with Regulations DD and E, which are similar across the five federal banking regulators, consist largely of a review of an institution's written policies and procedures and a sample of disclosure documents. Since 2005, NCUA has included examination procedures specifically addressing institutions' adherence to the 2005 interagency guidance concerning overdraft protection products and, in September 2007, all of the regulators revised their Regulation DD examination procedures to include reviews of the disclosures associated with such products offered by institutions that advertise them. While regulators received a large number of checking account complaints, they received relatively fewer complaints specifically concerning fees and related disclosures—less than 5 percent of all complaints received from 2002 to 2006. Further, the regulators reported a total of 1,674 instances in which they cited an institution for violation of the fee-related disclosure sections of Regulations DD and E from 2002 to 2006 (an average of about 335 annually among the nearly 17,000 institutions these regulators supervise). According to the regulators, the regulators took only two formal enforcement actions during this period related to these violations because most institutions took corrective actions during the course of the examination or shortly thereafter. The six selected state regulators we spoke with told us that their primary focus is on safety and soundness issues and compliance with state laws and regulations. Four of the six state regulators told us that they assess compliance with federal regulations such as Regulations DD and E. Like the federal regulators, the states reported receiving relatively few consumer complaints associated with checking and savings account fees and disclosures.

Our visits to 185 branches of depository institutions nationwide suggest that consumers shopping for accounts may find it difficult to obtain account terms and conditions and disclosures of fees upon request prior to opening an account. Similarly, our review of the Web sites of the banks, thrifts, and credit unions we visited suggests that this information may also not be readily available on the Internet. We were unable to obtain, upon request, a comprehensive list of all checking and savings account fees at 40 of the branches (22 percent) that we visited. Similarly, we were unable to obtain the account terms and conditions, including information on when deposited funds became available and how overdrafts were handled, for checking and savings accounts at 61 of the branches (33 percent). The results are consistent with those reported by a consumer group that conducted a similar exercise in 2001.[9] While the revised Regulation DD examination procedures call specifically for reviewing disclosures associated with overdraft protection products, the federal banking regulators do not have procedures to assess whether potential customers actually receive these or other disclosures. Consumers may consider convenience or other factors besides costs when shopping for checking or savings accounts, but this inability to obtain information about fees and the conditions under which fees are assessed upon request prior to opening a checking and savings account hinders their ability to make meaningful comparisons among institutions.

This report contains recommendations to the five federal banking regulators to incorporate into their supervision of financial institutions a means of ensuring that fee and other disclosure documents are made available to consumers upon request before opening an account, as intended by TISA and Regulation DD.

We requested and received written comments on a draft of this report from FDIC, the Federal Reserve, NCUA, OCC, and OTS that are presented in appendixes V through IX. In their written responses, all five banking regulators indicated agreement with our report and stated that they will be taking action in response to our recommendation. For example, OCC stated that it would incorporate steps, as needed, into its oversight of institutions' compliance with TISA to assure that disclosures continue to be made available. The Federal Reserve and NCUA specifically mentioned the need to revise, improve, or strengthen the current interagency Regulation DD examination procedures. All five agencies indicated that they plan to address this issue on an interagency basis. We also received technical comments from FDIC and the Federal Reserve, which we have incorporated in this report as appropriate.

BACKGROUND

Depository institutions—banks, thrifts, and credit unions—have attained a unique and central role in U.S. financial markets through their deposit-taking, lending, and other activities. Individuals have traditionally placed a substantial amount of their savings in federally insured depository institutions. In addition, the ability to accept deposits transferable by checks and other means has allowed depository institutions to become principal agents or middlemen in many financial transactions and in the nation's payment system. Depository institutions typically offer a variety of savings and checking accounts, such as ordinary savings, certificates of deposits, interest-bearing checking, and non-interest-bearing checking accounts. Also, the same institutions may offer credit cards, home equity lines of credit, real estate mortgage loans, mutual funds, and other financial products.

In the United States, regulation of depository institutions depends on the type of charter the institution chooses.[10] The various types of charters can be obtained at the state or national level and cover: (1) commercial banks, which originally focused on the banking needs of businesses but over time broadened their services; (2) thrifts, which include savings banks, savings associations, and savings and loans and which were originally created to serve the needs—particularly the mortgage needs—of those not served by commercial banks; and (3) credit unions, which are member-owned cooperatives run by member-elected boards with a historic emphasis on serving people of modest means.

All depository institutions have a primary federal regulator if their deposits are federally insured. State regulators participate in the regulation of institutions with state charters. Specifically, the five federal banking regulators charter and oversee the following types of depository institutions:

- OCC charters and supervises national banks. As of December 30, 2006, there were 1,715 commercial banks with national bank charters. These banks held the dominant share of bank assets, about $6.8 trillion.
- The Federal Reserve serves as the regulator for state-chartered banks that opt to be members of the Federal Reserve System and the primary federal regulator of bank holding companies, including financial holding companies.[11] As of December 30, 2006, the Federal Reserve supervised 902 state member banks with total assets of $1.4 trillion.
- FDIC supervises all other state-chartered commercial banks with federally insured deposits, as well as federally insured state savings banks. As of December 30, 2006, there were 4,785 state-chartered banks and 435 state-chartered savings banks with $1.8 trillion and $306 billion in total assets, respectively. In addition, FDIC has backup examination authority for federally insured banks and savings institutions of which it is not the primary regulator.
- OTS charters and supervises federally chartered savings associations and serves as the primary federal regulator for state-chartered savings associations and their holding companies. As of December 30, 2006, OTS supervised 761 federally chartered and 84 state chartered thrifts with combined assets of $1.4 trillion.
- NCUA charters, supervises, and insures federally chartered credit unions and is the primary federal regulator for federally insured state chartered credit unions. As of December 30, 2006, NCUA supervised 5,189 federally chartered and insured 3,173 state chartered credit unions with combined assets of $710 billion.

These federal regulators conduct on-site examinations and off-site monitoring to assess institutions' financial condition and compliance with federal banking and consumer laws. Additionally, as part of their oversight the regulators issue regulations, take enforcement actions, and close failed institutions.

Regulation DD, which implements TISA, became effective with mandatory compliance in June 1993. The purpose of the act and its implementing regulations is to enable consumers to make informed decisions about their accounts at depository institutions through the use of uniform disclosure documents. These disclosure documents are intended to help consumers "comparison shop" by providing information about fees, annual percentage yields, interest rates, and other terms for deposit accounts. The regulation is supplemented by "staff commentary," which contains official Federal Reserve staff interpretations of Regulation DD. Since the initial implementation date for Regulation DD, several amendments have been made to the regulation and the corresponding staff commentary. For example, the Federal Reserve made changes to Regulation DD, effective July 1, 2006, to address concerns about the uniformity and adequacy of information provided to consumers when they overdraw their deposit accounts.[12] Credit unions are governed by a substantially similar regulation issued by NCUA.[13]

Regulation E, which implements the Electronic Fund Transfer Act, became effective in May 1980. The primary objective of the act and Regulation E is the protection of individual consumers engaging in electronic funds transfers (EFT). Regulation E provides a basic framework that establishes the rights, liabilities, and responsibilities of participants in electronic fund transfer systems such as ATM transfers, telephone bill-payment services, point-of-sale terminal transfers in stores, and preauthorized transfers from or to consumer's bank accounts (such as direct deposit and Social Security payments). The term "electronic fund transfer" generally refers to a transaction initiated through an electronic terminal, telephone, computer, or magnetic tape that instructs a financial institution either to credit or to debit a consumer's asset account. Regulation E requires financial institutions to provide consumers with initial disclosures of the terms and conditions of EFT services. The regulation allows financial institutions to combine the disclosure information required by the regulation with that required by other laws such as TISA as long as the information is clear and understandable and is available in a written form that consumers can keep.

Paying or honoring customers' occasional or inadvertent overdrafts of their demand deposit accounts has long been an established practice at depository institutions. As shown in figure 1, depository institutions have four options when a customer attempts to withdraw or access funds from an account that does not have enough money in it to cover the transaction, and fees can be assessed for each of these options. The institution can (1) cover the amount of the overdraft by tapping a linked account (savings, money market, or credit card) established by the customer; (2) charge the overdraft to a linked line of credit; (3) approve the transaction (if electronic) or honor the customer's check by providing an ad hoc or "courtesy" overdraft; or (4) deny the transaction or decline to honor the customer's check. The first two options require that customers have created and linked to the primary checking account one or more other accounts or a line of credit in order to avoid overdrafts. The depository institution typically waives fees or may charge a small fee for transferring money into the primary account (a *transfer* fee). Depository institutions typically charge the same amount for a courtesy overdraft (an *overdraft* fee) as they do for denying a transaction for insufficient funds (an *insufficient funds* fee).

Institutions' options		Customer actively signs up for option	Type of fee
Overdraft to a linked account Institution pays overdraft by transferring funds from customer's linked account through an automated process.	Overdrawn account – Linked savings account, money market account, or credit card	✓	A per-transaction overdraft transfer fee.[a]
Overdraft to a line of credit Institution pays overdraft by charging customer's linked line of credit through an automated process.	Overdrawn account – Line of credit: Credit charged for amount of overdraft	✓	A per-transaction overdraft transfer fee.[b]
Ad hoc overdraft Institution's decision is discretionary and may be manual or automated, but an overdraft program is not publicized to the institution's customers.	Overdrawn account – Bank		A per-transaction overdraft fee.
"Courtesy" overdraft Institution's decision is discretionary and may be manual or automated, but the institution publicizes or promotes an overdraft program to its customers. The institution also typically discloses the dollar limit for covering overdrafts.	Overdrawn account – Bank		A per-transaction overdraft fee.
Transaction denied[c] Institution does not honor transaction, usually a check.			A per-transaction insufficient funds fee.

Source: GAO.

[a]Some banks may charge only one transfer fee per day. Also, if consumers link overdrafts to credit cards, then they may be subject to finance charges in addition to a transfer fee.

[b]The consumer may be subject to finance charges in addition to a transfer fee.

[c]If an electronic transaction is denied at the point of sale because of insufficient funds, the consumer typically is not charged an insufficient funds fee because the transaction is not completed. For payments involving checks, merchants may also charge a returned check fee in addition to what is charged by the bank.

Figure 1. Possible Outcomes of an Insufficient Funds Transaction.

In addition to fees associated with insufficient funds transactions, institutions may charge a number of other fees for checking and savings account services and transactions. As shown in table 1, these fees include periodic service charges associated with these accounts and special service fees assessed on a per-transaction basis.

Table 1. Selected Periodic and Special Service Fees Associated with a Checking or Savings Account

Fee	Applicability
Account maintenance	Assessed typically on a monthly basis for maintaining a checking or savings account. Depository institutions frequently waive routine service fees for customers who maintain a monthly minimum balance or meet other requirements, such as for direct deposits of paychecks.
Electronic banking or bill payment services	Assessed typically on a monthly basis for customers who opt for electronic banking or bill payment services.
ATM surcharge	Assessed by a depository institution for a non-account holder's use of its ATM.
Foreign ATM	Assessed on a transaction basis by a depository intuition for an account-holder's use of another depository institution's ATM.
Returns of deposited items	Assessed on a transaction basis by a depository institution when its account holder deposits a check that is then returned unpaid to the originating institution (for example, because of insufficient funds).
Stop payment order	Assessed by a depository institution for processing an account holder's order to withhold payment on a check already written. Source: GAO.

Some Fees on Checking and Savings Accounts Increased between 2000 and 2007, and Institutions' Reported Increasing Revenues from Fees

Our analysis of data from private vendors showed that a number of bank fees—notably charges for insufficient funds and overdraft transactions—have generally increased since 2000, while others have decreased.[14] In general, banks and thrifts charged higher fees than credit unions for checking and savings account services, and larger institutions charged more than smaller institutions. During this same period, the portion of depository institutions revenues derived from non-interest sources—including, but not limited to, fees on savings and checking accounts—increased somewhat. Changes in both consumer behavior and practices of depository institutions are likely influencing trends in fees, but limited data exist to demonstrate the effect of specific factors. FDIC is currently conducting a special study of the overdraft programs that should provide important insights on how these programs operate, as well as information on characteristics of customers who pay overdraft bank fees.

SINCE 2000, CHECKING AND SAVINGS ACCOUNT FEES HAVE INCREASED FOR SOME TRANSACTIONS AND SERVICES AND DECLINED FOR OTHERS

Data we obtained from vendors—based on annual surveys of hundreds of banks, thrifts, and credit unions on selected banking fees indicated that some checking and savings account fee amounts generally increased between 2000 and 2007, while a few fell, notably monthly maintenance fees.[15] For example, as shown in figure 2, average insufficient funds and overdraft fees have increased by about 11 percent, stop payment order fees by 17 percent, and return deposited item fees by 49 percent since 2000.[16]

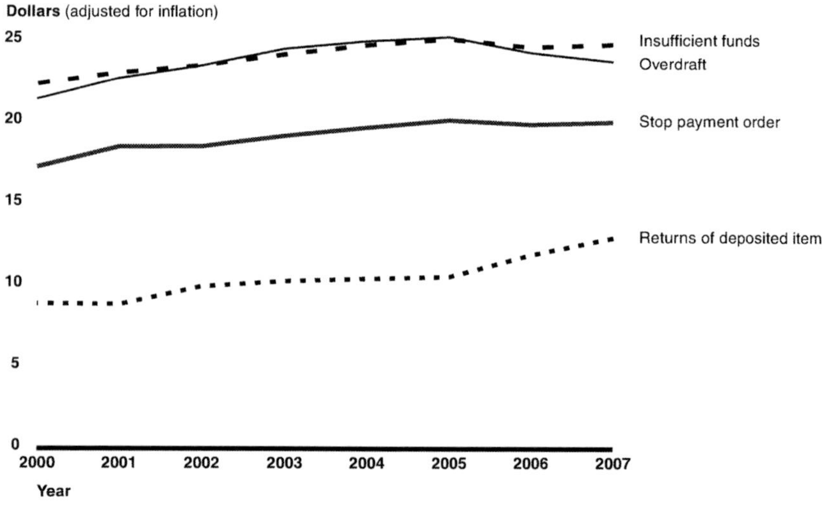

Source: GAO analysis of Moebs $ervices data.

Figure 2. Average Insufficient Funds, Overdraft, Return of Deposited Item, and Stop Payment Order Fees, All Institutions, 2000-2007.

Across all institutions, average insufficient funds and overdraft fees were the highest dollar amounts, on average, of the fees reported. For example, the average insufficient funds fee among the institutions surveyed by Moebs $ervices in 2006 was $24.02, while among the institutions surveyed by Informa Research Services it was $26.07. Data from Informa Research Services also indicated that since 2004 a small number of institutions (mainly large banks) have been applying tiered fees to certain transactions, such as overdrafts. For example, an institution may charge one amount for the first three overdrafts in a year (tier 1), a higher rate for overdrafts four to six of that year (tier 2), and an even higher rate for overdrafts seven and beyond in a single year (tier 3). Of the institutions that applied tiered fees in 2006, the average overdraft fees were $26.74, $32.53, and $34.74 for tiers 1, 2, and 3, respectively.

The data from these vendors also indicate that fee amounts for some transactions or services varied or generally declined during this period. For example:

- The average ATM surcharge fee (assessed by a depository institution when its ATM is used by a non-account holder) among institutions surveyed by Moebs $ervices was $0.95 in 2000, rising to $1.41 in 2003, and declining to $1.34 in 2006. This variability was also evident in the fees charged by institutions surveyed by Informa Research Services.
- The average foreign ATM fee (assessed by a depository institution when its account holders use another institution's ATM) generally declined, from $0.92 in 2000 to $0.61 in 2006 among institutions surveyed by Moebs $ervices and from $1.83 to $1.14 over the same period among institutions surveyed by Informa Research Services.
- The average monthly maintenance fees on standard non-interest bearing checking accounts decreased from $6.81 in 2000 to $5.41 in 2006 among institutions surveyed by Informa Research Services (Moebs $ervices did not provide data on this fee). Additionally, an increasing number of the surveyed institutions offered free checking accounts (with a minimum balance required to open the account) over this period. For example, in 2001 almost 30 percent of the institutions offered free checking accounts, while in 2006 the number grew to about 60 percent of institutions.

Finally, some fees declined in amount, as well as in terms of their prevalence. For example, Moebs $ervices reported that the institutions it surveyed charged annual ATM fees, generally for issuing a card to customers for their use strictly at ATMs, ranging from an average of $1.37 in 2000 to $1.14 in 2003. However, Moebs $ervices stopped collecting data on this fee because, according to a Moeb's official, fewer and fewer institutions reported charging the fee. Similarly, Moebs $ervices reported that the institutions it surveyed charged an annual debit card fee, generally for issuing a card to customers for their use at ATMs, averaging from $0.94 in

2000 to $1.00 in 2003; but, it stopped collecting this data as well. (Informa Research Services reported data on these fees through 2006, when they averaged $0.44 and $0.74, respectively.) Appendix III contains further details on the data reported by Moebs $ervices and Informa Research Services, in both nominal and real dollars.

A number of factors may explain why some fees increased while others decreased. For example, greater use of automation and lower cost of technology may explain why certain ATM fees have decreased or been eliminated altogether. Additionally, competition among depository institutions for customers likely has contributed to the decrease in monthly maintenance fees and the increased prevalence of "free checking" accounts. Factors that may be influencing trends in fees overall are discussed subsequently in this report.

FEES GENERALLY VARIED BY TYPE AND SIZE OF INSTITUTION

Using data supplied by the two vendors, we compared the fees for checking and savings accounts by type of institution and found that, on average, banks and thrifts charged more than credit unions for almost all of them (the exception was the fee for returns of deposited items).[17] For example, banks and thrifts charged on average roughly three dollars more than credit unions for insufficient funds and overdraft fees throughout the period. However, on average credit unions charged almost $6.00 more than banks and thrifts on returns of deposited items.

The amounts institutions charged for certain transactions also varied by the institution's size, as measured by assets. Large institutions—those with more than $1 billion in assets—on average charged more for the majority of fees than midsized or small institutions—those with assets of $100 million to $1 billion and less than $100 million, respectively. Large institutions on average charged between $4.00 and $5.00 more for insufficient funds and overdraft fees than smaller institutions. Further, on average, large banks and thrifts consistently charged the highest insufficient funds and overdraft fees, while small credit unions consistently charged the lowest. Specifically, in 2007 large banks and thrifts charged an average fee of about $28.00 for insufficient funds and overdraft fees, while small credit unions charged an average fee of around $22.00. While large institutions in general had higher fees than other sized institutions, smaller institutions charged considerably more for returns of deposited items. The results of our analysis are consistent with the Federal Reserve's 2003 report on bank fees, which showed that large institutions charged more than medium- and small-sized institutions (banks and thrifts combined) for most fees.[18]

Our analysis of Informa Research Services data also showed that, controlling for both institution type and size, institutions in some regions of the country, on average, charged more for some fees, such as insufficient funds and overdraft fees, than others. For example, in 2006 the average overdraft fee in the southern region was $28.18, compared with a national average of $26.74 and a western region average of $24.94.

Financial Institutions' Income from U.S. Government Accountability Office

INTEREST SOURCES, INCLUDING FEES, HAS INCREASED SINCE 2000

Between 2000 and 2006, the portion of depository institutions' income from non-interest sources, including income generated from bank fees, varied but generally increased. As shown in figure 3, banks' and thrifts' non-interest income rose from 24 to 27 percent of total income between 2000 and 2006 (peaking at 33 percent in 2004) and credit unions' non-interest income rose from 11 to 14 percent (peaking at 20 percent in 2004). The percent of non-interest income appeared to have an inverse relationship to changes in the federal funds rate—the interest rate at which depository institutions lend balances at the Federal Reserve to other depository institutions—which is an indicator of interest rate changes during the period. Low interest rates combined with increased competition from other lenders can make it difficult for banking institutions to generate revenues from interest rate "spreads," or differences between the interest rates that can be charged for loans and the rates paid to depositors and other sources of funds.

However, non-interest income includes revenue derived from a number of fee-based banking services, not all of them associated with checking and savings accounts. For example, fees from credit cards, as well as fees from mutual funds sales commissions, are included in non-interest income. Thus, non-interest income cannot be used to specifically identify either the extent of fee revenue being generated, or the portion that is attributable to any specific fee.

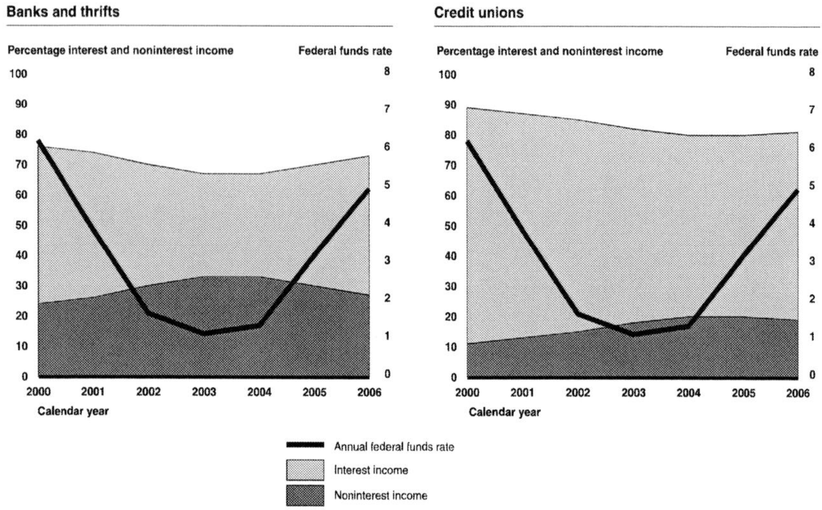

Sources: GAO analysis of FDIC's Statistics on Depository Institutions, NCUA's Financial Performance Report data, and the Federal Reserve's federal funds rate data.

Figure 3. Banks', Thrifts', and Credit Unions' Interest Income and Non-interest Income as a Percentage of Total Income and the Federal Funds Rate, 2000–2006.

Among other financial information, banks and thrifts are required to report data on service charges on deposit accounts (SCDA), which includes most of the fees associated with checking and deposit accounts.[19] Specifically, SCDA includes, among other things, account maintenance fees, charges for failing to maintain a minimum balance, some ATM fees, insufficient funds fees, and charges for stop payment orders. As figure 4 shows, banks' and thrifts' SCDA, and to a somewhat greater extent credit union's fee income as a percentage of total income, increased overall during the period, with a slight decline in recent years. However, it should be noted that credit union fee income includes income generated from both deposit accounts and other products that credit unions offer, such as fees for credit cards and non-customer use of proprietary ATMs; thus, the percentage of fee income they report is not directly comparable to the service charges reported by banks and thrifts.[20]

Because institutions do not have to report SCDA by line item, it is difficult to estimate the extent to which specific fees on checking and deposit accounts contributed to institutions' revenues or how these contributions have changed over the years. Further, some fees that banking customers incur may not be covered by SCDA. For example, institutions report monthly account maintenance fee income as SCDA, but not income earned from fees charged to a non-customer, such as fees for the use of its proprietary ATMs. Similarly, credit unions' reported fee income cannot be used to identify fee revenues from specific checking and savings account fees.

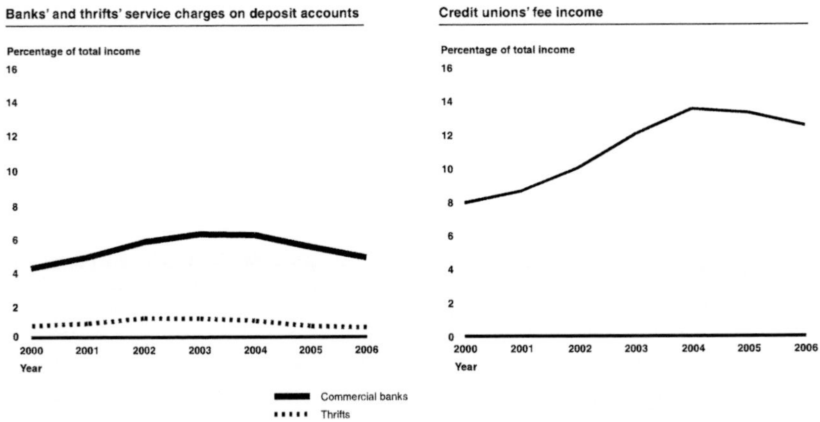

Sources: GAO analysis of FDIC's Statistics on Depository Institutions and NCUA's Financial Performance Report data.

Figure 4. Banks' and Thrifts' SCDA and Credit Unions' Fee Income as a Percentage of Total Income, 2000–2006.

Changes in Consumer Behavior and Depository Institution Practices May Affect Trends in Bank Fees

Since the mid-1990s, consumers have increasingly used electronic forms of payment such as debit cards for many transactions, from retail purchases to bill payment. By 2006 more than two-thirds of all U.S. non-cash payments were made by electronic payments (including credit cards, debit cards, automated clearing house, and electronic benefit transfers), while the number of paper payments (e.g., checks) has decreased due to the rapid growth in the use of debit cards.[21] Generally, these electronic payments are processed more quickly than traditional paper checks. For example, debit card transactions result in funds leaving customer's checking accounts during or shortly after the transaction, as opposed to checks, which may not be debited from a customer's account for a few days (although depository institutions have also begun to process checks faster, in part, as a result of the Check Clearing for the 21st Century Act (Check 21 Act) and implementing regulations, which became effective in late 2004).[22] Despite this overall shortening of time or “float” between the payment transaction and the debiting of funds from a consumer’s account, depository institutions can hold certain non-local checks deposited by a consumer for up to 11 days.[23] According to consumer groups and bank representatives, this creates the potential for increased incidences of overdrafts if funds are debited from a consumers account faster than deposits are made available for withdrawal. The shift in consumer payment preferences has occurred rather quickly, and we identified little research on the extent to which the increased use of electronic payments, such as debit cards, has affected the prevalence of specific deposit account fees, such as overdraft or insufficient fund fees.[24]

Additionally, some institutions have internal policies for posting deposits to and withdrawals from customer accounts that can affect the incidence of fees. For example, consumer group representatives, bank representatives, and federal regulatory officials told us that many institutions process the largest (highest dollar amount) debit transaction before the smallest one regardless of the order in which the customers initiated the transactions. This practice can affect the number of overdraft fees charged to a customer. For example, if a customer had only $600 available in their account, processing a payment for $590 first before three transactions of $25 each would result in three instances of overdrafts, whereas reversing the order of processing payments from smallest to largest would result in one instance of overdraft. Banking officials said that this processing of largest to smallest transactions first ensures that consumers’ larger, and presumably more important payments, such as mortgage payments, are made. One of the federal banking regulators—OTS—issued guidance in 2005 stating that institutions it regulates should not manipulate transaction clearing steps (including check clearing and batch debit processing) to inflate fees. We were unable to identify comprehensive information regarding the extent to which institutions were using this or other methods (chronological, smallest-to-largest, etc.) of processing payments.

Further, some depository institutions have automated the process used to approve overdrafts and have increasingly marketed the availability of overdraft protection programs to their customers. Historically, depository institutions have used their discretion to pay overdrafts for consumers, usually imposing a fee. Over the years, to reduce the costs of reviewing individual items, some institutions have established policies and automated the process for deciding whether to honor overdrafts, but generally institutions are not required to inform customers about internal policies for determining whether an item will be honored or denied. In addition, third-party vendors have developed and sold automated programs to institutions, particularly to smaller institutions, to handle overdrafts. According to the Federal Reserve, what distinguishes the vendor programs from in-house automated processes is the addition of marketing plans that appear designed to (1) promote the generation of fee income by disclosing to account holders the dollar amount that the consumer typically will be allowed to overdraw their account and (2) encourage consumers to use the service to meet short-term borrowing needs.[25] An FDIC official noted that some vendor contracts tied the vendor's compensation to an increase in the depository institution's fee revenues.

We were unable to identify information on the extent to which institutions were using automated overdraft programs developed and sold by third-party vendors or the criteria that these programs used. Representatives from a few large depository institutions told us that they are using software programs developed in-house to determine which account holders would have overdrafts approved. According to consumer groups and federal banking regulators, software vendors appear to be primarily marketing automated overdraft programs to small and midsized institutions. The 2005 interagency guidance on overdraft protection programs encouraged depository institutions to disclose to consumers how transactions would be processed and how fees would be assessed. An FDIC official noted that, while no empirical data are available, institutions' advertising of overdraft protection programs appears to have diminished since publication of the interagency guidance.

No Public Data Currently Exist on Characteristics of Consumers That Incur Bank Fees, but FDIC May be Able to Provide Some Information in the Future

Because fees for overdrafts and instances of insufficient funds may be more likely to occur in accounts with lower balances, there is some concern that they may be more likely among consumers who traditionally have the least financial means, such as young adults and low- and moderate-income households. We were not able to analyze the demographic characteristics of customers that incur bank fees because doing so would require transaction-level data for all account holders—data that are not publicly available. We identified only two studies—one by an academic researcher and one by a consumer group—that discussed the characteristics of consumers who pay bank fees. Neither study obtained a sample of customers who overdraw that was representative of the U.S. population. According to the academic researcher's study, which used transaction level account data for one small Midwest bank, overdrafts were not significantly correlated with consumers' income levels, although younger consumers were more likely to have overdrafts than consumers of other ages.[26] However, the results of this study cannot be generalized to the larger population because the small institution used was not statistically representative of all depository institutions. The consumer group study, which relied on a survey in which individuals with bank accounts were interviewed, found that those bank customers who had had two or more overdrafts in the 6 months before the date of the interview were more often low income, single, and non-white.[27] However, this study also had limitations, including the inherent difficulty in contacting and obtaining cooperation from a representative sample of U.S. households with a telephone survey and because it relied on consumers' recall of and willingness to accurately report past events rather than on actual reviews of their transactions. While we cannot fully assess the quality of results from these two studies, we note them here to illustrate the lack of definitive research in this area.

Partly in response to consumer concerns raised by overdraft protection products, FDIC is currently conducting a two-part study on overdraft protection products offered by the institutions it supervises. The results of this study may provide information on the types of consumers who pay bank fees. For both parts, FDIC is collecting data that are not currently available in the call reports or other standard regulatory reports. During the first phase of its study, FDIC collected data from 500 state-chartered non-member banks about their overdraft products and policies. Data from the first phase will reveal how many FDIC-regulated banks offer overdraft protection programs and the details of these programs, such as how many of them are automated. FDIC expects to complete the data collection effort at the end of 2007. The second phase involves collecting transaction-level data on the depositors who use the overdraft products for 100 of the 500 institutions for a year. As part of this phase, FDIC plans to use income information by U.S. Census Bureau tract data as a proxy for account holder's income to try and determine the characteristics of consumers who incur overdraft fees. FDIC expects to complete the analysis at the end of 2008.

REGULATORS FOCUS ON DEPOSITORY INSTITUTIONS' COMPLIANCE WITH FEDERAL DISCLOSURE REQUIREMENTS

Federal regulators assess depository institution's compliance with the disclosure requirements of Regulations DD and E during examinations by reviewing an institution's written policies and procedures, including a sample of disclosure documents. In general, regulators do not review the reasonableness of such fees unless there are safety and soundness concerns. Since 2005, NCUA has included examination procedures specifically addressing institutions' adherence to the 2005 interagency guidance concerning overdraft protection products and, in September 2007, all of the regulators revised their Regulation DD examination procedures to include reviews of the disclosures associated with such products offered by institutions that advertise them. In general, examinations are risk-based—that is, targeted to address factors that pose risks to the institution—and to help focus their examinations of individual institutions, the regulators review consumer complaints. Our analysis of complaint data from each of the federal regulators showed that while they receive a large number of checking account complaints, a small percentage of these complaints concerned the fees and disclosures associated with either checking or savings accounts. The federal regulators reported identifying a number of violations of the disclosure sections of Regulations DD and E during their examinations but collectively identified only two related formal enforcement actions from 2002 through 2006. Finally, officials from the six state regulators told us that, while they may look at compliance with Regulations DD and E, their primary focus is on safety and soundness issues and compliance with state laws and regulations, and they reported receiving few consumer complaints associated with checking and savings account fees and disclosure issues.

FEDERAL REGULATORS PRIMARILY REVIEW POLICIES, PROCEDURES, AND DISCLOSURE DOCUMENTS

Our review of the examination handbooks and examination reports indicated that the five federal regulators used similar procedures to assess compliance with Regulations DD and E (as discussed below, NCUA also includes steps to assess credit unions' adherence to the 2005 interagency guidance on overdraft protection products, but that is distinct from assessing compliance with regulatory requirements).[28] In general, the Regulation DD and E compliance examination procedures for each of the five federal banking regulators called for examiners to

- verify that the institution had policies or procedures in place to ensure compliance with all provisions of the regulations;
- review a sample of account disclosure documents and notices required by the regulation to determine whether contents were accurate and complete; and
- review a sample of the institution's advertisements to (1) determine if the advertisements were misleading, inaccurate, or misrepresented the deposit contract and (2) ensure that the advertisements included all required disclosures.

Federal regulators' examination procedures for Regulations DD and E do not require examiners to evaluate the reasonableness of fees associated with checking and savings accounts. According to the Federal Reserve, the statutes administered by the regulators do not specifically address the reasonableness of fees assessed. Additionally, officials of the federal regulators explained that there were no objective industry-wide standards to assess the "reasonableness" of fees.[29] OCC officials told us that an industry-wide standard would not work because, among other things, fees vary among banks that operate in different geographical areas and that competitive conditions in local markets determine fees. According to the federal regulatory officials, each depository institution is responsible for setting the fee for a particular product and service, and regulators look at rates or pricing issues only if there is a safety and soundness concern. For example, NCUA officials told us that an examiner's finding that fee income was excessive could create safety and soundness issues, depending on the way the fees were generated and how the resulting revenues were spent.

The regulators stated that while they did not evaluate the reasonableness of fees, the disclosure requirements of Regulations DD and E were intended to provide consumers with information that allow them to compare fees across institutions. Additionally, they told us that market forces should inhibit excessive fees since the financial institution would likely lose business if it decided to charge a fee that was significantly higher than its competitors.

Recent Revisions to Regulation DD Examination Procedures Require Further Review of Disclosures for Institutions Advertising Overdraft Protections

On September 13, 2007, the Federal Financial Institutions Examination Council's Task Force on Consumer Compliance—a formal interagency body composed of representatives of the Federal Reserve, FDIC, NCUA, OCC, and OTS—approved revised interagency compliance examination procedures for Regulation DD. Officials of each of the federal regulators told us that their agencies either had begun or were in the process of implementing the updated examination procedures. Among other changes, the revised examination procedures address the Regulation DD disclosure requirements for institutions that advertise the payment of overdrafts. Specifically, the revised examination procedures ask the examiners to determine whether the institution clearly and conspicuously discloses in its advertisements (1) the fee for the payment of each overdraft, (2) the categories of transactions for which a fee may be imposed for paying an overdraft, (3) the time period by which a consumer must repay or cover any overdraft, and (4) the circumstances under which the institution will not pay an overdraft.

These items are among those that were identified as "best practices" by the 2005 interagency guidance. According to the guidance, clear disclosures and explanations to consumers about the operation, costs, and limitations of an overdraft protection program are fundamental to using such protection responsibly. Furthermore, the guidance states that clear disclosures and appropriate management oversight can minimize potential customer confusion and complaints, as well as foster good customer relations. The interagency guidance identifies best practices currently observed in or recommended by the industry on marketing, communications with consumers, and program features and operations. For example, the best practices include marketing the program in a way that does not encourage routine overdrafts, clearly explaining the discretionary nature of the program, and providing the opportunity for consumers to opt out of the program.

Prior to the revised Regulation DD examination procedures, NCUA had adopted procedures to assess the extent to which institutions it examines followed the interagency guidance. In December 2005, NCUA adopted "bounce protection" (that is, overdraft protection) examination procedures as part of the agency's risk-focused examination program. The examination procedures were developed to coincide with the issuance of the 2005 interagency guidance on overdraft protection programs, according to an NCUA official.[30] In an NCUA letter to credit unions, the agency stated that "credit unions should be aware the best practices are minimum expectations for the operation of bounce protection programs."[31] NCUA's examination procedures included a review of several key best practices. For example, the examination procedures assess whether credit unions provided customers with the opportunity to elect overdraft protection services or, if enrollment in such a program was automatic, to opt out. In addition to other areas of review, the examination procedures include a review of whether the credit union distinguished overdraft protection from "free" account features, and if the credit union clearly disclosed the fees of its overdraft protection program.

To a more limited extent, OTS had overdraft protection examination procedures in place that address its guidance, but these were limited to a review of compliance-related employee training and the materials used to market or educate customers about the institution's overdraft protection programs. Officials from the Federal Reserve, OCC, and FDIC reported that, beyond the recent revisions to Regulation DD examination procedures, their agencies did not have specific examination procedures related to the 2005 interagency guidance because the best practices are not enforceable by law. These officials told us that, while not following a best practice from the interagency guidance did not constitute a violation of related laws or regulations, they encourage institutions to follow the best practices. An FDIC official noted that a deviation from the guidance could serve as a "red flag" for an examiner to look more closely for potential violations.

While Federal Regulators Received a Large Number of Checking Account Complaints, a Small Percentage Were Related to Fees and Disclosures

Officials of the federal banking regulators explained that examiners use complaint data to help focus examinations that they are planning or to alter examinations already in progress. For example, according to one regulator, if consumers file complaints because they have not received a disclosure document prior to opening an account, this could signify a violation of Regulation DD, which the examiners would review as part of the examination for this regulation. The officials noted that consumer complaints could be filed and were often resolved at the financial institution involved, in which case the consumer would not be likely to contact a federal banking regulator.[32] However, if the consumer is not satisfied with the financial institution's response, a consumer would then likely file a complaint with the federal banking regulator. Consumers may also file a complaint directly with federal regulators without contacting the financial institution about a problem. In either case, regulators are required to monitor the situation until the complaint is resolved.[33]

According to the regulators' complaint data, most of the complaints received from 2002 to 2006 involved credit cards, although a significant number of complaints were related to checking accounts and a somewhat smaller number involved savings accounts (fig. 5). In analyzing complaints specifically about checking and savings accounts from 2002 through 2006, we found that, on average, about 10 percent were related to fees, and 3 percent were related to disclosures. (For information on how the Federal Reserve, FDIC, OCC, and OTS resolved complaints, see app. IV.) Collectively fee and disclosure complaints represented less than 5 percent of all complaints received during this period. Officials of the banking regulators told us that the overwhelming bulk of complaints they received on checking and saving accounts concerned a variety of other issues, including problems opening or closing an account, false advertising, and discrimination.

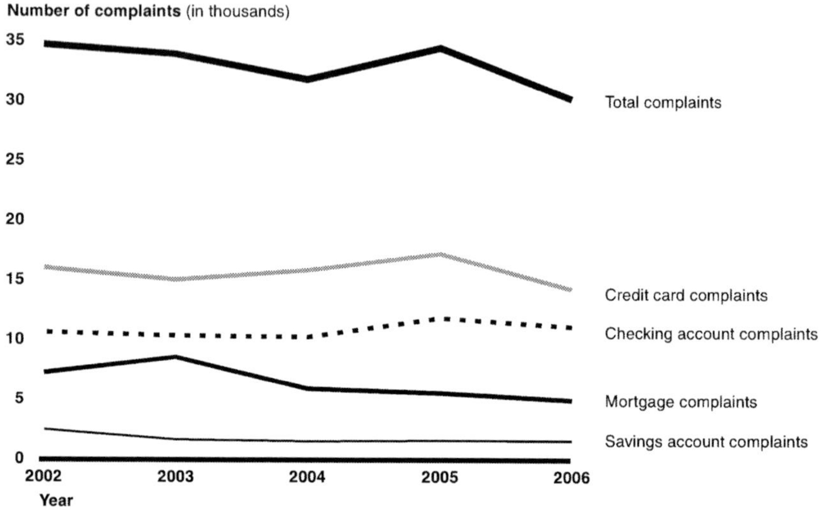

Sources: GAO analysis of OCC, OTS, NCUA, FDIC, and Federal Reserve data.

Note: For the combined period of 2002 to 2006, over 70 percent of the complaints were filed against national banks, which are supervised by OCC.

Figure 5. Complaints Related to Four Major Products for All Federal Regulators.

Among the regulators, OCC included in its complaint data the specific part of the regulation that was the subject of the complaint. Of the consumer complaints about fees that OCC received from 2002 through 2006, 39 percent were for "unfair" fees (concerning the conditions under which fees were applied), 2 percent were for new fees, 6 percent were for "high" fees (the amount of the fees), and 53 percent concerned fees in general. The majority of disclosure-related complaints that OCC received during this period were for the Regulation DD provision that, in part, requires that depository institutions provide account disclosures to a consumer before an account is opened or a service is provided, whichever is earlier, or upon request. OCC's analysis of these complaints serves to identify potential problems—at a particular bank or in a particular segment of the industry—that may warrant further investigation by examination teams, supervisory guidance to address emerging problems, or enforcement action.

Federal Regulators Identified a Number of Violations of Fee-Related Disclosure Provisions during Their Examinations but Took Few Related Enforcement Actions

The federal banking regulators' examination data for the most recent 5 calendar years (2002 through 2006) showed a total of 1,674 instances in which the regulators cited depository institutions for non-compliance with the fee-related disclosure requirements of Regulations DD (1,206 cases) or E (468 cases). On average, this is about 335 instances annually among the nearly 17,000 depository institutions that these regulators oversee. As shown in table 2, most of the disclosure-related violations were reported by FDIC—83 percent of the Regulation DD disclosure-related violations (998 of 1,206) and 74 percent of the Regulation E disclosure-related violations (348 of 468). According to FDIC officials, one reason for the larger number of fee-related violations identified by FDIC is the large number of institutions for which it is the primary federal regulator (5,220 depository institutions as of December 31, 2006). Also, differences among the regulators may appear due to the fact that they do not count the numbers of violations in exactly the same way.

Table 2. Number of Regulation DD and E Disclosure-Related Violations Identified by Federal Banking Regulators from 2002-2006

Regulator	Number of Regulation DD disclosure-related violations	Number of Regulation E disclosure-related violations	Number of compliance examinations between 2000–2006[a]	Number of institutions regulated in 2006
FDIC	998	348	9,876	5,220
Federal Reserve	67	99	1,526	902
NCUA	102	10	35,757[b]	8,362
OCC	26	8	6,566	1,715
OTS	11	3	3,203	761
Total	**1,206**	**468**	**43,282**	**16,960**

Sources: GAO analysis of FDIC, Federal Reserve, NCUA, OCC, and OTS data.

Note: The fee–related disclosure violations represent cited instances of noncompliance with sections 230.3(a), 230.3(b), 230.4(a) and 230.4(b)(4) of Regulation DD and sections 205.4(a)(1), 205.7(a), and 205.7(b)(5) of Regulation E.

[a]Examinations are risk focused, and not all examinations assess compliance with Regulations DD and E.

[b]This number represents the number of federally chartered and state-chartered institutions examined.

According to our analysis of the regulators' data, the most frequent violation associated with the initial disclosure requirements of Regulation DD was non-compliance with the requirement that disclosure documents be written in a clear and conspicuous manner, in a form that customers can keep, and reflect the terms of the legal obligation of the account agreement between the consumer and the depository institution (1,053 cases). Examiners reported violations of two other disclosure provisions of Regulation DD. First, they found violations of the requirement that depository institutions provide account disclosure documents to a consumer before an account is opened or a service is provided, whichever is earlier, or upon request (124 cases). Second, they reported violations of the requirement that disclosure documents state the amount of any fee that may be imposed in connection with the account or an explanation of how the fee will be determined and the conditions under which it may be imposed (29 cases).

The most frequent violation associated with the initial disclosure requirements of Regulation E was of the requirement that financial institutions make the disclosure documents available at the time a consumer contracts for an EFT or before the first EFT is made involving the consumer's account (321 cases). Other disclosure provisions from Regulation E for which examiners cited violations included those that required disclosure statements to be in writing, clear and readily understandable, and in a form that customers can keep (5 cases) and to list any fees imposed by the financial institution for EFTs or for the right to make transfers (142 cases).

According to officials of the federal banking regulators, examiners are typically successful in getting the financial institutions to take corrective action on violations either during the course of the examination or shortly thereafter, negating the need to take formal enforcement action. FDIC, NCUA, OCC, and Federal Reserve officials reported that from 2002 to 2006 they had not taken any formal enforcement actions solely related to violations of the disclosure requirements from Regulations DD and E, while OTS reported taking two such actions during the period.[34]

STATE REGULATORS RELIED ON AND WORKED WITH FEDERAL REGULATORS TO REVIEW REGULATIONS DD AND E AND REPORTED FEW CONSUMER COMPLAINTS ABOUT FEES AND DISCLOSURES

Officials of all six of the state banking regulators that we contacted told us that the primary focus of their examinations is on safety and soundness issues and compliance with state laws and regulations. Officials of four of the six state banking regulators we contacted told us their examiners also assess compliance with Regulation DD, and three of these four indicated that they assess compliance with Regulation E as well. Representatives of the four state banking regulators also told us that if they identify a violation and no federal regulator is present, they cite the institution and forward this information to the appropriate federal banking regulator. The other two state banking regulators said that they review compliance with federal regulations, including Regulations DD and E, only if the federal banking regulators have identified non-compliance with federal regulations during the prior examination.

Officials in four states said that their state laws and regulations contained additional fee and disclosure requirements beyond those contained in Regulations DD and E. For example, according to Massachusetts state banking officials, Massachusetts bank examiners review state-chartered institutions for compliance with a state requirement that caps the fees on returns of deposited items. In another example, an Illinois law restricts institutions from charging an ATM fee on debit transactions made with an electronic benefits card (a card that beneficiaries used to access federal or state benefits, such as food stamp payments), according to Illinois state banking officials. Additionally, these state officials told us that Illinois state law requires all state-chartered institutions to annually disclose their fee schedules for consumer deposit accounts. According to an official at the New York state banking department, their state has a number of statutes and regulations concerning bank fees and their disclosure to consumers and their state examiners review institutions' compliance with these requirements. The laws and regulations cover, among other things, permissible fees, required disclosure documents, and maximum insufficient fund fees, according to the New York state officials.

Two of the states reported that, in conducting examinations jointly with the federal regulators, they had found violations of the Regulation DD and E disclosure provisions from 2002 to 2006 (one state reported 1 violation of

Regulation DD, and one state reported 16 violations of Regulation DD and 10 violations of Regulation E). Four of the states did not report any violations (in one case, the state agency reported that they did not collect data on violations). Three states also reported that they had not taken any formal enforcement actions against institutions for violations of Regulation DD or E disclosure provisions; two states reported that they did not collect data on enforcement actions for violations of these regulations; one state did not report any data to us on enforcement actions. Regarding consumer complaints, officials in two states said that they did not maintain complaint data concerning fees and disclosures associated with checking and savings accounts, and the other four states reported relatively few complaints associated with fees and disclosures. For example, Massachusetts reported a total of 89 complaints related to fees and disclosures during the period, in comparison to 4,022 total complaints over the period.

Despite Federal Regulations and Compliance Examinations, We Experienced Difficulty Obtaining Fee Information

The results of our requests for information on fees or account terms and conditions at depository institutions we visited, as well as our visits to institutions' Web sites, suggest that consumers may find it difficult to obtain such information upon request prior to opening a checking or savings account. A number of factors could explain the difficulties we encountered in obtaining comprehensive information on fees and account terms and conditions, including branch staff potentially not being knowledgeable about federal disclosure requirements or their institution's available disclosure documents. Further, federal banking regulators' examination processes do not assess whether potential customers can easily obtain information that institutions are required to disclose. Potential customers unable to obtain such information upon request prior to opening an account will not be in a position to make meaningful comparisons among institutions, including the amounts of fees they may face or the conditions under which fees would be charged.

Federal Laws and Regulations Require Disclosures so That Consumers Can Make Meaningful Comparisons among Institutions

As we have seen, TISA requires, among other things, that depository institutions provide consumers with clear and uniform disclosures of the fees that can be assessed against all deposit accounts, including checking and savings accounts, so that consumers may make a meaningful comparison between different institutions. Depository institutions must provide these disclosures to consumers before they open accounts or receive a service from the institution or upon a consumer's request. Regulation DD and the accompanying staff commentary specify the types of information that should be contained in these disclosures, including

- minimum balance required to open an account;
- monthly maintenance fees and the balance required to avoid them;
- fees charged when a consumer opens or closes an account;
- fees related to deposits or withdrawals, such as charges for using the institution's ATMs; and
- fees for special services—for example, insufficient funds or charges for overdrafts and stop payment order fees on checks that have been written but not cashed.

Regulation DD also requires depository institutions to disclose generally the conditions under which a fee may be imposed—that is, account terms and conditions. For example, institutions must specify the categories of transactions for which an overdraft fee may be imposed but do not have to provide an exhaustive list of such transactions.

While depository institutions are required to provide consumers with clear and uniform disclosures of fees to enable meaningful comparisons among institutions, consumers may consider other factors when shopping among institutions. For example, federal banking regulators and one consumer group told us that convenience factors, such as locations of branches or ATMs, are typically the factors that consumers consider the most besides costs, when choosing where to open a checking and savings account.

We Encountered Difficulties in Obtaining a Comprehensive List of Fees or Terms and Conditions during Our Visits to Depository Institution Branches and Web Sites

Our visits to branches of depository institutions nationwide suggested that some consumers may be unable to obtain, upon request, meaningful information with which to compare an institution's fees and how they are assessed before opening a checking or savings account. We also found that the institutions' Web sites generally did not provide comprehensive information on fees or account terms and conditions. Further, the documents that we did obtain during our visits did not always describe some key features of the institutions' internal policies and procedures that could affect the incidence or amount of overdraft fees assessed by the institution.

Information Was Not Always Available during Visits to Depository Institution Branches

To assess the ease or difficulty in obtaining a comprehensive list of fees and account terms and conditions associated with checking and savings accounts, GAO staff from 12 cities across the United States visited 185 branches of banks, thrifts, and credit unions.[35] Collectively, these branches represented 154 different depository institutions. Posing as potential customers, we specifically requested a comprehensive list of fees and terms and conditions for checking and savings accounts that would allow us to compare such information across depository institutions.[36] The results are summarized here.

- *Comprehensive list of fees.* We were unable to obtain a comprehensive list of fees for checking and savings accounts from 40 (22 percent) of the branches (representing 36 institutions). Instead, we obtained brochures describing only the features of different types of checking and savings accounts. Some of these brochures contained information on monthly maintenance fees and the minimum balance needed to avoid them. But these brochures did not contain information on other fees, such as overdraft or insufficient fund fees.

While our success in obtaining a comprehensive list of fees varied slightly among institutions of different sizes, we did note greater variations among banks, credit unions, and thrifts. For example, we were unable to obtain a comprehensive list of fees at 18 percent of the 103 bank branches and 20 percent of the 46 credit union branches we visited (representing 14 banks and 9 credit unions, respectively), while among the 36 thrift branches visited (representing 13 thrift institutions) it was 36 percent.[37]

- *Account terms and conditions.* We were unable to obtain the terms and conditions associated with checking and savings accounts from 61 of the 185 branches (representing 54 depository institutions) that we visited (33 percent). Instead, as described earlier, we were provided with brochures on the different types of checking and savings accounts offered by the institution.

  We also observed little differences in our ability to obtain account terms and conditions information from institutions of different sizes but again found differences by types of institutions. For example, we were unable to obtain this information at 32 percent of the small or midsized institutions (34 of 108), compared with 35 percent of the large institutions (27 of 77). With respect to the type of depository institution, we were unable to obtain these documents at 30 percent of the bank branches (31 of 103 branches, representing 25 banks), 35 percent of the credit union branches (16 of 46 branches, representing 16 credit unions), and 39 percent of the thrift branches (14 of 36 branches, representing 13 thrift institutions).

For both the comprehensive list of fees and descriptions of account terms and conditions, we observed some differences among branches of a single depository institution. For example, we visited multiple branches of 23 depository institutions (that is, more than one branch of each of the 23). For four of these institutions, we were able to obtain all of the documents we requested from all of the branches. For the other 19 institutions, we encountered inconsistencies among the different branches in our ability to obtain the full set of information we requested.

The results of our direct observations are generally consistent with those reported by the U.S. Public Interest Research Group (PIRG). In 2001, PIRG had its staff pose as consumers and visit banks to request fee brochures and reported that, in many cases, its staff members were unable to obtain this information despite repeated requests.[38] Further, our results seem to be in accord with the violations data provided by the regulators; as noted previously, the most frequent violation of the fee-related disclosure provisions of Regulation DD cited by the regulators between 2002 and 2006 was non-compliance with the requirement that disclosure documents be written in a clear and conspicuous manner and in a form that customers can keep.

Information Was Not Available on Many Institutions' Web Sites

While depository institutions are not required to have the comprehensive list of fees and account terms and conditions on Web sites if these sites are merely advertising and do not allow consumers to open an account online, we visited these Web sites as part of our effort to simulate a consumer trying to obtain information to compare checking and savings accounts across institutions. In visiting the Web sites of all the institutions that we visited in person, we were unable to obtain information on fees and account terms and conditions at more than half of them. For example, we were unable to obtain a comprehensive list of fees from 103 of the 202 Web sites (51 percent). In addition, we were unable to obtain the terms and conditions from 134 of the 202 (66 percent). Figure 6 compares the results of our visits to branches and Web sites of depository institutions.

Some of the depository institutions' Web sites nevertheless contained information on certain fees associated with checking and savings accounts. For example, most of the Web sites had information on monthly maintenance fees and ATM fees associated with checking accounts. Smaller percentages had information on fees for overdrafts and insufficient fund fees. For example,

- 87 percent provided information on monthly maintenance fees,
- 62 percent had information on ATM withdrawal fees,
- 41 percent contained information on overdraft fees, and
- 37 percent provided information on insufficient fund fees.

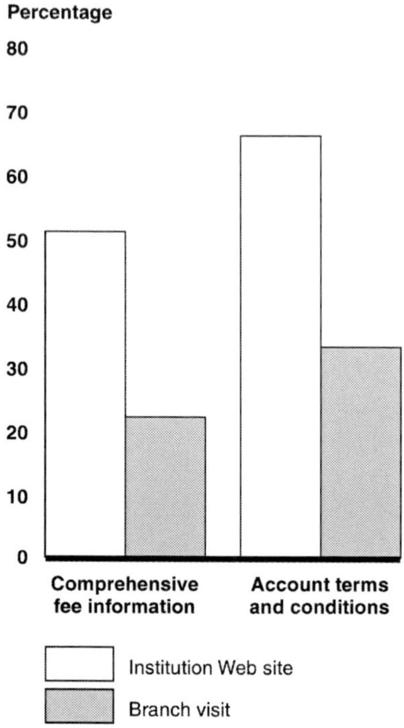

Source: GAO.

Figure 6. Percentage of Depository Institution Branches and Web Sites We Visited That Did Not Provide a Comprehensive List of Fees and Terms and Conditions.

SEVERAL FACTORS COULD EXPLAIN WHY WE HAD DIFFICULTY OBTAINING INFORMATION CONCERNING FEES

Among branches at which we were unable to obtain a comprehensive list of fees, branch staff offered explanations suggesting that they may not be knowledgeable about federal disclosure requirements. As previously noted, depository institutions are required to provide consumers, upon request, with clear and uniform disclosures of the fees that can be assessed against checking and savings accounts so that consumers may make a meaningful comparison between different institutions. However, during our visits to branches of depository institutions,

- representatives at 14 branches we visited told us that we had all the information on fees we needed to comparison shop—even though we determined that the documents they provided did not include a comprehensive list of fees that consumers opening accounts there might have to pay,
- representatives at seven branches told us that no comprehensive fee schedules were available, and
- representatives at four branches told us that we had to provide personal information or open an account in order to obtain a comprehensive list of fees.

In addition, we observed differences in our ability to obtain the comprehensive list of fees and account terms and conditions among branches of 19 of the 23 depository institutions we visited that had multiple branches. This variation among branches of the same institution suggests that staff knowledge of the institution's available disclosure documents may have varied.

Further, the examination procedures that federal banking regulators use to assess compliance with Regulation DD do not require examiners to verify whether new or potential customers are actually able to obtain the required disclosure documents before opening an account. (Rather, the examination procedures call for the examiner to review written policies and procedures and disclosure documents to ensure that they contain information required under the regulation). As a result, examination results would not provide officials of depository institutions with information showing whether potential customers were experiencing difficulty obtaining information at particular branches.

Because the results of our visits cannot be generalized to other institutions, and because the federal banking regulators do not assess the extent to which consumers are actually able to obtain disclosure documents, neither we nor the regulators know how widespread this problem may be, nor—to the extent that it does exist among institutions—the reasons for it. However, regardless of the cause, if consumers are unable to obtain key information upon request prior to opening an account, they will be unable to make meaningful distinctions regarding charges and terms of checking and savings accounts.

CONCLUSIONS

The amounts of some fees associated with checking and savings accounts have grown over the past few years, while others have varied or declined. During the same time period, the portion of depository institutions' incomes derived from non-interest sources, including fees, has varied somewhat but has risen overall. Changes in both consumer behavior, such as increased use of electronic forms of payment, and in the terms and conditions of accounts offered by depository institutions may be influencing these trends in fees, but available data do not permit determining their exact effects. Similarly, we could find little information on the characteristics of consumers who are most likely to incur fees. However, the general upward trend in fees puts a premium on the effective disclosure of account terms and conditions, including the amounts of individual fees and the conditions under which they will be assessed, to consumers who are shopping for savings and deposit accounts.

While consumers may consider convenience or other factors, as well as costs, when choosing a depository institution, Regulation DD, as well as guidance issued by the federal banking regulators, is intended to ensure that consumers receive information needed to make meaningful comparisons among institutions regarding the savings and deposit accounts they offer. While the federal regulators take consumer complaints into account when determining the scope of their examinations of specific institutions, their examinations of compliance with Regulations DD and E consist of reviewing institutions' written policies, procedures, and disclosure documents. On this basis, the regulators have cited numbers of institutions for violating the disclosure requirements. Further, the regulators are in the process of implementing revised examination procedures for Regulation DD compliance that will include assessing the extent to which depository institutions follow requirements governing the advertisement of overdraft protection programs. This will be particularly important given that fees associated with overdrafts were among the highest of the types of fees for which we obtained data. However, even under the revised procedures, the regulators' examinations do not determine whether consumers actually receive required disclosure documents before opening an account.

While the results of our visits to 185 branches of depository institutions cannot be generalized to all institutions, they raise some concern that consumers may find it difficult to obtain upon request, important disclosure documents prior to opening an account. We were unable to obtain detailed information about fees and account terms and conditions at over one-fifth of the branches we visited and, in many cases, we found inconsistencies among branches of the same depository institution. Because the federal banking regulators, in their compliance examinations, do not assess the extent to which consumers actually receive required disclosure documents before opening an account, they are not in a position to know how widespread this problem may be among the institutions they supervise, or the reasons for it. Incorporating into their oversight a means of assessing the extent to which consumers can actually obtain information to make meaningful comparisons among institutions, and taking any needed steps to assure the continued availability of such information, would further this goal of TISA.

Recommendations for Executive Action

To help ensure that consumers can make meaningful comparisons between depository institutions—we recommend that the Chairman, Federal Deposit Insurance Corporation; Chairman, Board of Governors of the Federal Reserve System; Chairman, National Credit Union Administration; Comptroller of the Currency, Office of the Comptroller of the Currency; and Director, Office of Thrift Supervision assess the extent to which consumers receive specific disclosure documents on fees and account terms and conditions associated with demand and deposit accounts prior to opening an account, and incorporate steps as needed into their oversight of institutions' compliance with TISA to assure that disclosures continue to be made available.

Appendix I. Objectives, Scope, and Methodology

Our report objectives were to determine (1) the trends in the types and amounts of fees associated with checking and deposit accounts since 2000; (2) how federal and selected state banking regulators address checking and deposit account fees in their oversight of depository institutions; and (3) the extent to which consumers are able to obtain account terms and conditions and disclosures of fees, including information about specific transactions and bank practices that determine when such fees are assessed, upon request prior to opening an account.

Available Data on the Types and Amounts of Bank Fees and Characteristics of Consumers Who Incur These Fees

To provide information on the average amounts of various checking and savings account fees, we purchased data from two market research firms that specialize in the financial services industry; Moebs $ervices and Informa Research Services. Moebs $ervices provided us with an electronic file that contained data from 2000 to 2007 on the following fees:

- annual automated teller machine (ATM) fees,
- ATM surcharges,

- foreign ATM fees,
- insufficient funds fees,
- overdraft fees,
- overdraft transfer fees from a line of credit,
- overdraft transfer fees from a deposit account,
- return deposited item fees,
- stop payment order fees and
- debit card annual fees.

Moebs $ervices collected its data through telephone surveys with financial service personnel at each sampled institution. In the surveys, callers used a "mystery shopping" approach and requested rates and fees while posing as potential customers. The surveys were completed in June for each of the years we requested (the 2006 survey was conducted in December), and we obtained data from the following number of institutions (table 3):

Table 3. Number of Institutions Surveyed by Moebs $ervices, 2000-2007

	Year							
	2000	**2001**	**2002**	**2003**	**2004**	**2005**	**2006**	**2007**
Institutions	2,955	3,937	3,937	4,396	5,838	5,261	5,264	5,492

Source: GAO analysis of Moebs $ervices data.

Note: The number of institutions sampled each year varied because of adding core based statistical areas, expanding number of states surveyed, a decrease in overall institutions, and refining the accuracy and precision of the sample selection.

The statistical design of the survey was developed for Moebs $ervices by Professor George Easton of Emory University. The design consisted of a stratified random sample by (1) institution type (banks and thrifts combined, and credit unions), (2) institution size (as shown in table 4), and (3) regions of the country defined by metropolitan statistical area.

Table 4. Definition of Institution Size Categories

Institution size	Asset size
Small institutions	Assets less than $100 million
Midsized institutions	Assets between $100 million to $1 billion
Large institutions	Assets more than $1 billion

Source: GAO analysis of Moebs $ervices data.

We took the data we obtained from Moebs $ervices and computed average fees for institutions overall, as well as for institutions by type, size, and region.

We interviewed Moebs $ervices representatives to understand their methodology for collecting the data and ensuring its integrity. In addition, we conducted reasonableness checks on the data we received and identified any missing, erroneous, or outlying data. We also worked with Moebs $ervices representatives to ensure our analysis of their data was correct. Finally, for the years 2000 through 2002, we compared the average fee amounts we calculated with averages the Board of Governors of the Federal Reserve System (Federal Reserve) had calculated using Moebs $ervices data for their "Annual Report to the Congress on Retail Fees and Services of Depository Institutions." We found our averages to be comparable to those derived by the Federal Reserve and determined that the Moebs $ervice's data were reliable for the purposes of this report.

Informa Research Services also provided us with an electronic file that included summary level fee data from 2000 to 2006. The data included information for the same fees that Moebs $ervices had provided, but, also included the following fees:

- monthly fees for checking and savings account;
- insufficient funds and overdraft tiered fees;
- check enclosure and imaging fees;
- foreign ATM balance inquiry fees; and
- foreign ATM denied transaction fees.

In addition to fee data, Informa Research Services also provided us with data on the minimum balances required to open an account, the monthly balances needed to waive fees, and the maximum number of overdrafts or insufficient funds fees that an institution would charge per day. Informa Research Services collected its data by gathering the proprietary fee statements of the financial institutions, as well as making anonymous in-branch, telephone, and Web site inquiries for a variety of bank fees. Informa Research Services also receives the information directly from its contacts at the financial institutions. The data are not statistically representative of the entire population of depository institutions in the country because the company collects fee data for particular institutions in specific geographical markets so that these institutions can compare their fees against their competitors. That is, surveyed institutions are self-selected into the sample, or are selected at the request of subscribers. To the extent that institutions selected in this manner differ from those which are not, results of the survey would not accurately reflect the industry as a whole. Informa Research Services collects data on over 1,500 institutions, including a mix of banks, thrifts, credit unions, and Internet-only banks. The institutions from which it collects data tend to be large institutions that have a large percentage of the deposits in a particular market. Additionally, the company has access to individuals and information from the 100 largest commercial banks. Table 5 shows the mix of institutions for which Informa Research Services collected fee type data from 2000–2006.

Table 5. Number of Institutions for Which Informa Research Services Collected Data, 2000–2006

	Year						
	2000	**2001**	**2002**	**2003**	**2004**	**2005**	**2006**
Institutions	119	308	412	947	1,150	1,418	1,571

Source: GAO analysis of Informa Research Services data.

The summary level data Informa Research Services provided us for each data element included the average amount, the standard deviation, the minimum and maximum values, and the number of institutions for which data were available to calculate the averages. Informa Research Services also provided this summary level data by the same categories of institution type and size as the Moebs $ervices data. In addition, Informa Research Services provided us with data for nine specific geographic areas: California, Eastern United States, Florida, Michigan, Midwestern United States, New York, Southern United States, Texas, and Western United States.

We interviewed Informa Research Services representatives to gain an understanding of their methodology for collecting the data and the processes they had in place to ensure the integrity of the data. We also conducted reasonableness checks on the data and identified any missing, erroneous, or outlying data and worked with Informa Research Services representatives to correct any mistakes we found. As we did with the Moebs $ervices data, we compared the average fee amounts Informa Research Services had calculated for selected fees for 2000, 2001, and 2002 with the Federal Reserve's "Annual Report to the Congress on Retail Fees and Services of Depository Institutions." We found the averages to be comparable to those derived by the Federal Reserve and determined that the Informa Research Services data were sufficiently reliable for this report. To evaluate bank fee trends, for both the Moebs $ervices and Informa Research Services data, we adjusted the numbers for inflation to remove the effect of changes in prices. The inflation adjusted estimates used a base year of 2006 and Consumer Price Index calendar year values as the deflator.

To determine the extent to which bank fees are contributing to depository institutions' revenue, we obtained data from the quarterly financial information (call reports) filed by depository institutions and maintained by the Federal Deposit Insurance Corporation (FDIC). From this data, we analyzed interest income, non-interest income, and service charges on deposit accounts for commercial banks and thrifts from 2000 to 2006. We analyzed the data for all institutions, as well as by institution type (banks versus thrifts) and institution size (assets greater than $1 billion, assets between $100 million and $1 billion, and assets less than $100 million). Similarly, for credit unions, we reviewed the National Credit Union Administration's (NCUA) "Financial Performance Reports," which provided quarterly data for interest income, non-interest income, and fee income for all federally insured credit unions from 2000 to 2006. Based on past work, we have found the quarterly financial data maintained by FDIC and NCUA to be sufficiently reliable for the purposes of our reports.

To determine the effect, if any, of changing consumer payment preferences and bank processing practices on the types and frequency of account fees incurred by consumers, we reviewed the 2004 and 2007 Federal Reserve payment studies on non-cash payment trends in the United States.[1] We also reviewed data on payment trends in debit and credit card transactions from the *EFT Data Book*.[2] In addition, we spoke with multiple industry experts, including bank representatives and consumer group representatives, such as the Consumer Federation of America, the Center for Responsible Lending, and the U.S. Public Interest Research Group to understand what practices banks employ to process transactions on deposit accounts, how these practices have changed over the past few years, and the potential impact these practices have had on consumers incurring fees, such as overdraft fees. Furthermore, we reviewed studies that analyzed electronic payment preferences and identified one study that used transaction-level data to determine how payment preferences influence overdraft fees.[3]

To determine what data are available on the characteristics of consumers who pay bank fees, we reviewed two studies on the topic; one by an academic researcher and another by a consumer group. The academic study used transaction-level account data and regression models to estimate the probability of overdrawing an account. The data included customer information and all transactions with associated balances from May-August 2003, from one small Midwestern bank.[4] The second study used data collected by telephone surveys of 3,310 adults, who were 18 years or older, between October 2005 and January 2006.[5] Both studies suffer from limitations that preclude making inferences to the broader populations of banking customers who pay fees, but they represent the only relevant research at this point, and are suggestive of the characteristics of these customers. We also reviewed documentation on and interviewed officials at the FDIC about their ongoing study of overdraft protection programs, including the phase of their study in which they will review transaction-level data. Finally, we interviewed two academic researchers and representatives of eight consumer groups; five depository institutions; two software vendors; and four industry trade associations, including the American Bankers Association, Independent Community Bankers of America, America's Community Bankers, and the Credit Union National Association, to determine what research had been done on the topic.

How Regulators Address Fees Associated with Checking and Deposit Accounts

To assess the extent that federal and selected state banking regulators review fees associated with checking and deposit accounts as part of their oversight of depository institutions, we obtained and reviewed examination manuals and guidance used by the five federal banking regulators—Federal Reserve, FDIC, NCUA, the Office of the Comptroller of the Currency (OCC), and the Office of Thrift Supervision (OTS)—and conducted interviews with agency officials. We also obtained and reviewed a sample of 25 compliance examination reports, on examinations completed during 2006, to identify how the federal regulators carried out examinations for compliance with Regulations DD and E. We selected five examination reports from each regulator based on an institution's asset size and geographic dispersion, in an attempt to capture a variety of examinations. The asset size of the institutions ranged from $2 million to $1.2 trillion. In addition, we obtained information on the regulatory efforts of six states. We selected the states based on recommendations from the Conferences of State Banking Supervisors, New York State Banking Department, and Massachusetts Division of Banks and to achieve geographical dispersion. The selected states were: California, Connecticut, Illinois, Maine, Massachusetts, and New York. We reviewed compliance examination manuals and guidance used by the six state regulators and asked specific questions to each state's appropriate banking officials.

To determine the number of complaints that the regulators received on checking and savings accounts, in addition to complaints about fees and disclosures, we requested complaint data, including data on resolutions, for calendar years 2002 through 2006. For the complaint data, we obtained data on the banking products or services involved, the complaint category and, in some cases, the citation of the regulation. While our estimates of the proportions of complaints related to fees depend on how the banking regulators coded the subjects of the complaint they received, and how we combined those related to fees, we judge any possible variations to be slight. For the complaint resolution data, we obtained information about the resolution (outcomes) of complaints and the banking products or services involved. The data came from five different databases: (1) OCC's REMEDY database, (2) the Federal Reserve's Complaint Analysis Evaluation System and Reports (CAESAR), (3) FDIC's Specialized Tracking and Reporting System (STARS), (4) OTS' Consumer Complaint System (CCS), and (5) NCUA's regionally based system on complaints. We obtained data from OCC, the Federal Reserve, FDIC, OTS, and NCUA that covered calendar years 2002 through 2006. For purposes of this report, we used data from the regulators' consumer complaint databases to describe the number of complaints that each regulator received related to fees and disclosures for checking and savings accounts, as well as complaints received by four major product categories—checking accounts, savings accounts, mortgage loans, and credit cards. With respect to the data on complaint resolutions, we used the regulators' data to describe the number of cases each regulator handled, what products consumers complained about, and how the regulators resolved the complaints. To assess the reliability of data from the five databases, we reviewed relevant documentation and interviewed agency officials. We also had the agencies produce the queries or data extracts they used to generate the data we requested. Also, we reviewed the related queries, data extracts, and the output for logical consistency. We determined these data to be sufficiently reliable for use in our report.

Finally, we obtained data from each of the federal regulators on violations they cited against institutions for non-compliance with Regulation DD and Regulation E provisions. Specifically, we asked for data on the total number of violations that each regulator cited for all examined provisions of Regulations DD and E during 2002 to 2006, as well as for data on violations of selected disclosure provisions. The Regulation DD sections that we requested and obtained data on were: §§ 230.3, 230.4, 230.8, and 230.11. The Regulation E sections that we requested and obtained data on were: §§ 205.4 and 205.7. We compiled the data and summarized the total number of violations found for all of the federal regulators during 2002 to 2006. We also obtained data from 2002 through 2006 on the total number of enforcement actions that each regulator took against institutions for violations of all provisions of Regulations DD and E and the selected disclosure provisions. To assess the reliability of data from the five databases, we reviewed relevant documentation and interviewed agency officials. We also had the agencies produce the queries or data extracts they used to generate the data we requested. Also, we reviewed the related queries, data extracts, and the output for logical consistency. We determined these data to be sufficiently reliable for use in our report. Finally, we also requested information from each state regulator on consumer complaint, violation, and enforcement data pertaining to bank fees and disclosures, state specific bank examination processes, and any additional state laws pertaining to bank fees and disclosures. We did not receive all our requested data because some states' systems did not capture complaint, violation, or enforcement data related to bank fees and disclosures. For those states where information was available, the number of complaints and violations were minimal and not consistently reported among states. We, therefore, attributed the limited information on complaints, violations, and enforcement actions to state officials and did not assess the reliability of this data.

Effective Disclosures of Fees to Consumers

To assess the extent to which consumers, upon request prior to opening a checking and savings account, are provided disclosures of fees and the conditions under which these fees are assessed, GAO employees visited 103 bank branches, 36 thrift branches, and 46 credit union branches of 154 depository institutions throughout the nation. We selected these institutions to ensure a mix of institution type (bank, thrift, and credit union) and size; however, the results cannot be generalized to all institutions. We reviewed the federal Truth-in-Savings Act (TISA) and Regulation DD, which implements TISA, to determine what disclosure documents depository institutions were required to provide to new and potential customers. Using a standardized, prescribed script, GAO employees posed as consumers and specifically requested a comprehensive fee schedule and terms and conditions associated with checking and savings accounts. The branches were located in the following cities: Atlanta, Georgia; Boston, Massachusetts; Chicago, Illinois; Dallas, Texas; Dayton, Ohio; Denver,

Colorado; Huntsville, Alabama; Los Angeles, California; Norfolk, Virginia; San Francisco, California; Seattle, Washington; and Washington, D.C.

The GAO employees visiting these branches also reviewed the institutions' Web sites to determine if these sites had comprehensive fee schedules and terms and conditions associated with checking and savings accounts. After both visiting branches and reviewing Web sites, GAO employees used standardized forms and recorded whether or not they were able to obtain the specific documents (examples were provided) and whether or not they were able to locate specific information on each institutions' Web site.

To obtain information on issues related to providing consumers with realtime account information during debit card transactions at point-of-sale terminals and automated teller machines (see app. II), we reviewed available literature from the Federal Reserve, including a 2004 report on the issues in providing consumers point-of-sale debit card fees during a transaction.[6] We also reviewed other sources that described the payment processing system related to debit card transactions at merchants and ATMs. In addition, we conducted structured interviews with officials from five banks, two card associations, three third-party processors, four bank industry associations, and one merchant trade organization, and summarized our findings.

We conducted this performance audit in Atlanta, Georgia; Boston, Massachusetts; Chicago, Illinois; Dallas, Texas; Dayton, Ohio; Denver, Colorado; Huntsville, Alabama; Los Angeles, California; Norfolk, Virginia; San Francisco, California; Seattle, Washington; and Washington, D.C., from January 2007 to January 2008 in accordance with generally accepted government auditing standards. Those standards require that we plan and perform the audit to obtain sufficient, appropriate evidence to provide a reasonable basis for our findings and conclusions based on our audit objectives. We believe that the evidence obtained provides a reasonable basis for our findings and conclusions based on our audit objectives.

Appendix II. Issues with Providing Consumers Real-Time Account Information at Point-of-Sale Terminals and ATMs When Using a Debit Card

According to debit card industry representatives we contacted, providing consumers with their "real-time" account balance information during a debit card transaction is technically feasible but presents a number of issues that would need resolution.[1] These issues include the costs associated with upgrading merchant terminals and software to allow for consumers' account balances to be displayed at the terminals; the potential difficulty of determining a consumer's real-time account balance, given the different types of transactions that occur throughout the day; concerns over privacy and security raised by account balances potentially being visible to others besides account holders; and the increased time it would take to complete a transaction at merchant locations.[2]

Challenges Stem in Part From Existing Technology and Steps Used to Accomplish Transactions

A consumer using a debit card to make a purchase at a merchant's checkout counter (referred to as a point-of-sale debit transaction) has two options for completing the transaction: (1) entering a personal identification number (PIN) or (2) signing for the transaction (similar to a credit card transaction). The consumer is typically prompted at the point-of-sale terminal to choose either "debit" (in which case the transaction is referred to as "PIN-based") or "credit" (in which case the transaction is referred to as "signature-based"). Regardless of which option the consumer chooses, the transaction is a debit card transaction. PIN- and signature-based debit card transactions differ not only with respect to the input required from the consumer but also the debit networks over which the transactions are carried and the number and timing of steps involved in carrying out the transactions. Similarly, transactions initiated at ATMs can differ in how they are processed. Customers can make withdrawals and deposits not only at ATMs owned by their card-issuing institutions but also at ATMs owned by other depository institutions or entities.[3] An ATM card is typically a dual ATM/debit card that can be used for both ATM and debit card transactions, both PIN-based and signature-based, at participating retailers. [4]

PIN-Based Debit Card Transactions

PIN-based debit card transactions are referred to as "single message" because the authorization—the approval to complete the transaction— and settlement—the process of transmitting and reconciling payment orders—of the transaction take place using a single electronic message.[5] As shown in figure 7, PIN-based debit card transactions involve a number of steps between the merchant's terminal and the consumer's deposit account. Generally, at the locations of large national merchants, after the consumer has swiped the card a message about the transaction is transmitted directly to the electronic funds transfer (EFT) network.[6] (For other merchants, the transaction reaches the EFT network via the merchant's processor, also known as the merchant acquirer.[7]) The message identifies the consumer's institution and account, the merchant, and the dollar amount of the purchase.

The EFT network routes the transaction to the card issuer (or to the card issuer's processor, which then passes it to the card issuer). The card issuer—usually the consumer's depository institution—receives the message and uses the identifying information to verify that the account is valid, that the card has not been reported lost or stolen, and that there are either sufficient funds available in the account or the account is covered by an overdraft protection program (that is, the issuer covers the transaction even if there are insufficient funds in the account, which is also known as bounce protection).[8] If these conditions are met, the issuer authorizes the debit transaction. Specifically, the issuer then debits the consumer's account and sends an authorization message to the EFT network, which sends it to the merchant's acquirer, which forwards the authorization to the merchant's terminal. The entire sequence typically occurs in a matter of seconds.

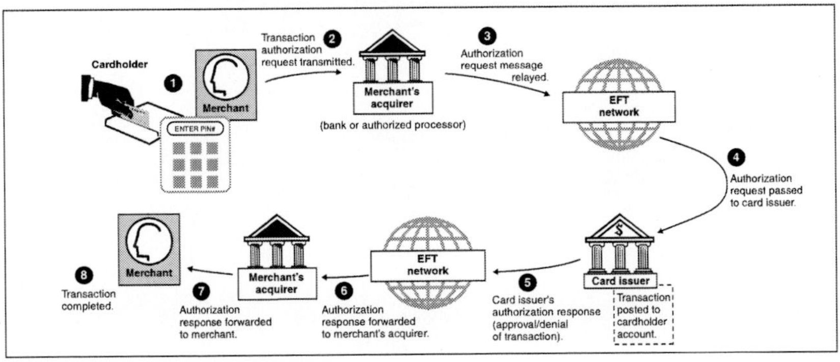

Sources: GAO and VISA; Art Explosion (clip art).

Figure 7. Path of a Typical PIN-Based Debit Card Transaction.

SIGNATURE-BASED DEBIT CARD TRANSACTIONS

Signature-based debit card transactions involve two electronic messages: one to authorize the transaction and another to settle the transaction between the merchant and the card issuer, at which time the consumer's account is debited. To conduct a signature-based debit card transaction, the customer typically has a VISA- or MasterCard-branded debit card linked to a deposit account. As shown in figure 8, after the card is swiped, a message about the transaction travels directly (or indirectly, through the merchant's acquirer) to the VISA or MasterCard network, from which the transaction proceeds directly (or indirectly, through the card-issuing institution's processor) to the card-issuing institution. As in a PIN-based debit card transaction, if the issuer verifies the relevant information, it authorizes the transaction and routes it back through the VISA or MasterCard network to the merchant's acquirer with the authorization. The merchant acquirer then forwards the authorization to the merchant's terminal, and the consumer signs the receipt.[9] The settlement of the transaction between the merchant and card issuer (and the actual debiting of the consumer's account) occurs after a second message is sent from the merchant to the issuer, usually at the end of the day.[10]

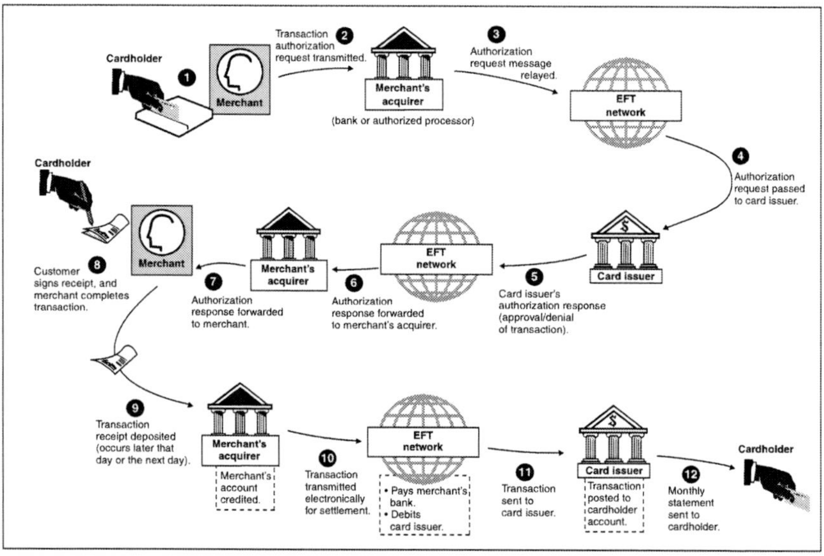

Sources: GAO and VISA; Art Explosion (clip art).

Figure 8. Path of a Typical Signature-Based Debit Card Transaction.

ATM-Based Transactions

The steps involved in ATM transactions depend upon whether a consumer is using an ATM owned by the issuer of his or her card (typically referred to as a "proprietary" ATM), or an ATM owned by a depository institution or entity other than the card-issuing institution (typically referred in the industry as a "foreign ATM"). A foreign ATM transaction is processed in essentially the same manner as a PIN-based debit card transaction, with one exception: the ATM operator (or its processor) routes the transaction to the EFT network, which then routes it to the card issuer.[11] The card-issuing institution authorizes the transaction via the EFT or debit card networks. In contrast, when a consumer uses a proprietary ATM, the transaction stays within the issuer's network and does not require the use of an external EFT network (fig. 9). Card issuers that are depository institutions—such as banks—may have the capability of providing a notice to their customers at a proprietary ATM that a withdrawal will result in the account being overdrawn and then allow the customer to decide whether or not to proceed with the transaction. Officials from one of the banks that we spoke with stated that they employed this capability at their proprietary ATMs.

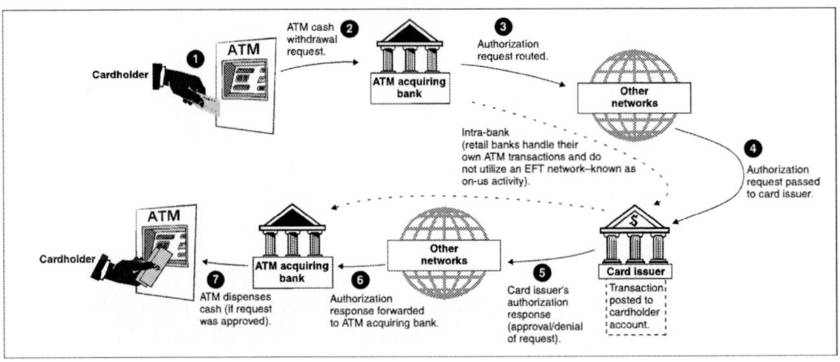

Sources: GAO and VISA; Art Explosion (clip art).

Figure 9. Path of a Typical Debit Card Transaction at an ATM.

Point-of-Sale Terminal and Software Upgrade Challenges

As of March 2007, there were over 5 million point-of-sale terminals in the United States.[12] According to industry representatives, most point-of-sale terminals are not currently equipped to display a consumer's checking account balance and, in these cases, merchants would either need to replace the terminal entirely or upgrade the software in the terminal. Industry representatives were hesitant to estimate the costs associated with this because the number of terminals that would need to be replaced versus those that would only need a software upgrade is not currently known. The industry representatives explained that the cost of upgrading the point-of-sale terminals to display account balance information would be primarily borne by merchants.

In addition to upgrading point-of-sales terminals, industry representatives identified the following other costs that would be incurred:

- Upgrading software used by the EFT networks and depository institutions in order to transmit balance information from the card-issuing institution to the merchant. As described above, currently a debit card transaction is authorized by verifying a consumer's checking account balance and sending back an approval or denial message—which does not include account balance information.
- Increasing the communications infrastructure of the EFT networks to allow for additional message traffic, namely consumers' acceptances or declinations of a transaction once they have viewed their account balances. These messages would constitute a second message from the point-of-sale terminal to the card-issuing institution for each transaction.

An associated cost with this process would be training employees who work at the terminals how to handle these debit card transactions and the cost of additional time to accomplish transactions, which we discuss here.

With respect to providing account balance information at foreign ATMs, one industry representative explained that this would require all entities involved in ATM transactions (banks, ATM operators, ATM networks, and the Federal Reserve) to agree on a common message format to display balances, as well as a new transaction set for ATMs that would provide consumers with the option not to proceed with the transaction once they saw their balances. Two industry representatives we spoke with said that it could take a number of years for all of the entities involved in ATM transactions to agree on a standard format.

Challenges in Providing Accurate Real-Time Account Balance Information

Debit card industry representatives explained that the account balance that is used to authorize a debit card transaction—and which would be displayed to the consumer—may not necessarily reflect the true balance in the consumer's checking account at the time of the transaction. One of the reasons for this is that, while a depository institution may attempt to get as close to a real-time balance as possible, it may be unable to capture all of the transactions associated with the account as they occur throughout the day. For example, one depository institution official told us that it updates its customers' account balances throughout each day; it refers to these updated balances as a customer's "available balance." This available balance is updated throughout the day to reflect debit card transactions at point-of-sale terminals and ATMs, as well as other transactions such as those that occur online. This balance, however, might not take into account checks that will be clearing that day, deposits made at a foreign ATM, or some transactions that would come in via the Automated Clearing House (ACH).[13] An example of the latter is a transaction in which a consumer electronically transfers funds from a mutual fund to a checking account. The net result of the inability to provide consumers with a real-time balance is that the consumer may be presented with a balance that is not reflective of all the transactions that will be processed as of that day.

Another reason why a depository institution may be unable to provide consumers with a real-time balance is that the institution may not update balances throughout the day. Most institutions "batch process" transactions at night, then post the revised customer account balances. The following day, the institutions update the customer's account balance for debit card authorizations and certain other transactions that occur throughout the day. However, according to a card association, some small banks only post the account balance from the batch process to the customer's account and do not update account balances as transactions occur throughout the day.

Finally, if a depository institution uses a third-party processor to authorize debit card transactions, the balance that the third-party processor uses may also not reflect all the transactions that occur throughout the day. For example, transactions involving a bank teller, such as deposits or withdrawals, do not require a third-party processor to authorize transactions, thus the processor would not be able to update its balance to reflect these transactions.

Challenges Associated with Privacy and Security Issues

One of the major concerns raised by the debit card industry representatives we spoke with regarding providing consumers with real-time balances at point-of-sale terminals was a concern over privacy. Unlike ATM transactions, which are transactions between a consumer and the machine, under which consumers tend to be cognizant of the need for privacy, point-of-sale terminals are generally more visible to others, according to these representatives. For example, the balance on a point-of-sale terminal could be visible to the cashier and customers in line at a merchant location. In addition, at restaurants, the waiter or other staff could view this information out of sight of the consumer. The industry representatives stated that most consumers would likely be uncomfortable having their account balance information visible to others.

Another related concern raised by these representatives was one of security, in that cashiers or possibly other customers might be able to view a consumer's account balance. Thus, the industry representatives stated that providing balances at a point-of-sale terminal could increase the risk of fraud. One industry representative told us that providing a balance at a point-of-sale terminal would be a departure from current privacy and security approaches with point-of-sale transactions.

Challenges Related to Increased Transaction Processing Time

Industry representatives explained that allowing consumers to accept or decline a transaction once they have viewed their balance would likely increase the time it takes to get customers through a check-out line. According to a retail merchant's trade association that we contacted, merchants depend on moving customers quickly through check-out lines. The retail merchants' trade association stated that adding a step in the check-out process would add time, resulting in lower sales volume per unit of time for each cashier, and potentially greater costs associated with adding cashiers to maintain the same volume of transactions.

Industry officials also stated that there were some circumstances during a point-of-sale transaction for which providing consumers with real-time balances would not be possible or would be problematic. For example, during "stand-in" situations, such as when a card issuer's systems are offline for maintenance, EFT networks review and authorize (or deny) transactions in accordance with instructions from the issuer. The networks would not have real-time access to account balance information when the issuer's system is down. Another example would be merchants, such as fast food outlets, who perform quick swipes of debit cards for low dollar transactions. At the time of the swipe, the merchant has not actually routed the transaction to the card issuer and thus has not yet accessed the consumer's account balance. In these cases, the merchant has accepted the risk of not being paid if there are insufficient funds in the account in order to move customers through lines more quickly. Finally, one industry representative questioned how the industry would be able to provide consumers with real-time balances if consumers make debit card purchases online or over the telephone.

Other Options Short of Providing Real-Time Account Balance Information at Point-of-Sale Terminals or ATMs Have Their Challenges and Limitations

There are other options short of providing real-time account balances at point-of-sale terminals and ATMs that might assist in warning consumers of a potential overdraft, but each of these options has challenges and limitations. For example, one option involves sending a warning with the authorization message instead of a real-time balance. The warning would indicate that the transaction could result in an overdraft. As indicated above, one of the banks we met with currently provides a similar warning on its proprietary ATMs. The consumer would then have the option to accept or deny the transaction. This option would require two messages to complete a debit card transaction rather than one message. Further, under this option, depository institutions would still be unable to base their authorization decisions on a real-time balance because of the various types of transactions that may occur in a day, and thus no warning message would be triggered—yet once the institution reconciles all accounts, a consumer could be faced with an overdraft fee. This option would also likely slow down transactions and raise costs for merchants. However, unlike providing real-time account balance information at a point-of-sale terminal, this option would not present privacy or security concerns because the balance in the consumer's account would not be transmitted.

Another option short of providing consumers with real-time account balance information is printing a consumer's available balance on a receipt after a transaction has been completed. This is currently possible when consumers use their card issuers' proprietary ATMs and some foreign ATMs, according to industry representatives. Under this option, the consumer would not receive a warning that the transaction could subject them to an overdraft, and they would not have a choice to accept or decline the transaction. Further, under this option, the consumer would not be provided his or her account balance until after the transaction was completed.[14] However, once consumers obtained their balance, they could change their spending behavior to avoid a fee on subsequent transactions. This option would entail certain costs for upgrading terminals or software in order to print the consumer's real-time balance on the receipt, as well as costs of upgrading software to transmit the real-time balance from the card-issuing institution to the merchant terminal. The option would not address an institution's ability to provide an actual real-time balance and would introduce privacy and security concerns because if the receipt were inadvertently dropped, others could view the balance. However, this option would not slow down the time it takes to complete a transaction because the consumer would not be given the option of accepting or declining a transaction.

Finally, industry representatives noted that consumers currently have a number of ways to check their account balances (e.g., by phone and Internet), which might help them avoid overdraft fees. According to Federal Reserve officials, this would require "near-time" processing and a system that synchronizes the balance information reported through the phone and Internet banking systems with the balance information that is transmitted by the institution to the ATM/EFT network. Three of the four large banks we spoke with stated that their customers currently have the ability to sign up for a feature in which the bank will send a message to the consumers' e-mail accounts or cell phones—"E-alerts"—when their balances reach a designated "threshold" amount. Under this option, consumers receiving an E-alert could change their spending patterns to avoid incurring an overdraft situation and fees. Table 6 compares the E-alert option with other potential options for warning consumers that they may incur an overdraft fee and the associated issues surrounding the particular option.

Table 6. Issues Raised by Options for Warning Consumers That They May Incur an Overdraft When Using a Debit Card at a Point-of-Sale Terminal or ATM

Options	Issues				
	Merchant hardware upgrades	**Software or communications infrastructure upgrades**	**Real-time balance accuracy**	**Privacy and security concerns**	**Speed of transaction**
• Disclosure of real-time balance at point-of-sale terminal or ATM	• Replace or reprogram terminals to display real-time balance	• Reprogram to allow balances to be sent with authorization message • Increase communications infrastructure to allow for two messages to complete each transaction	• Unable to ensure real-time balance	• Potential increase in risk of fraud if balance viewed by cashier or others	• Transaction slowed to allow consumer to accept or decline
• Notice of warning of overdraft and fee associated with overdraft disclosed at point-of-sale terminal or ATM	• Replace or reprogram terminals to display warning	• Reprogram to allow warning to be sent with authorization message • Increase communications infrastructure to allow for two messages to complete each transaction	• Unable to ensure real-time balance	• None	• Transaction slowed to allow consumer to accept or decline
• Printing of available balance on receipt after transaction completed at point-of-sale terminal or ATM	• Replace or reprogram terminals to print real-time balance	• Reprogram to allow balances to be sent with authorization message	• Unable to ensure real-time balance	• Potential increase in risk of fraud if receipt is lost, then found by others	• None
• E-alert to consumers' e-mail account or cell phone that their available balance is low	• None	• Program to allow alerts to be sent to e-mail or cell phone	• Unable to ensure real-time balance	• None	• None

Source: GAO.

APPENDIX III. ANALYSES OF SELECT BANK FEES DATA

Using the methodology we noted earlier, we analyzed select bank fee data obtained from two firms, Moebs $ervices and Informa Research Services. Some bank fees have increased since 2000, while a few, such as monthly fees, have decreased.[1] As noted earlier in the report, we analyzed data in aggregate for all depository institutions and also by institution type and size. According to data we obtained, banks and thrifts charged more than credit unions for almost all select fees analyzed, and larger institutions charged higher fees than midsized and smaller institutions. We found slight variations in fees charged by region, with certain regions charging less than the national average for some select bank fees analyzed. For example, California and the Western United States consistently charged less than the national average for almost all select fees analyzed according to the Informa Research Services data.

For both the Moebs $ervices and Informa Research Services data, banks and thrifts were combined into one institution type category, with credit unions as the other institution type. For both sets of data, the following asset size categories were used:

- small institutions had assets less than $100 million,
- midsized institutions had assets between $100 million and $1 billion, and

- large institutions had assets greater than $1 billion.

For the Moebs $ervices data, we computed average amounts ourselves, but statistics were provided to us for the Informa Research Services data. We identified all instances in which the information presented was based on data provided by less than 30 institutions and did not include those instances in this report because averages based on a small number of institutions may be unreliable.[2] The information presented for the Moebs $ervices data is statistically representative of the entire banking and credit union industry, but the Informa Research Services data is not. For additional information on the select fees analyzed and the number of institutions surveyed, see appendix I.

Analysis of Moebs $ervices Data

Table 7 provides a detailed comparison of the Moebs $ervices data for all institutions for select bank fees for the 8-year period, 2000–2007.

Table 7. Average Fees, All Institutions, 2000–2007

	Year							
Type of fee	**2000**	**2001**	**2002**	**2003**	**2004**	**2005**	**2006**	**2007**
Insufficient funds	$21.77	$22.43	$22.90	$23.57	$24.15	$24.52	$24.02	$24.18
Overdraft	20.83	22.10	22.88	23.93	24.40	24.65	23.69	23.13
Returns of deposited items	8.28	8.24	9.34	9.66	9.81	9.93	11.28	12.31
Stop payment order	16.66	17.91	17.93	18.58	19.09	19.53	19.28	19.41
Foreign ATM	0.92	0.96	0.81	0.79	0.83	0.84	0.61	0.69
ATM surcharge	0.95	1.23	1.29	1.41	1.37	1.40	1.34	1.29
Overdraft transfer fee (link to deposit account)	Data not available	Data not available	Data not available	2.75	2.86	2.91	2.81	2.69

Table 7. Continued

=	Year							
Type of fee	**2000**	**2001**	**2002**	**2003**	**2004**	**2005**	**2006**	**2007**
Overdraft transfer fee (link to line of credit)	Data not available	Data not available	Data not available	1.23	1.37	1.10	1.86	1.49
ATM annual fee	1.37	0.92	1.32	1.14	Data not Available	Data not available	Data not available	Data not available
Debit card annual fee	Data not available	$0.94	$0.87	$1.00	Data not available	Data not available	Data not available	Data not available

Source: GAO analysis of Moebs $ervices data.

Note: Fees are adjusted for inflation and reported in 2006 dollars.

ANALYSIS OF INFORMA RESEARCH SERVICES DATA

Table 8 provides a detailed comparison of the Informa Research Services data for all institutions for select bank fees for the 7-year period, 2000–2006. The data is presented for a variety of types of checking and savings accounts.[3]

Table 8. Average Fees, All Institutions, 2000–2006

	Year						
Type of fee	**2000**	**2001**	**2002**	**2003**	**2004**	**2005**	**2006**
Insufficient funds	Data not available	$27.78	$27.91	$26.94	$26.77	$26.56	$26.07
Overdraft	Data not available	28.31	28.19	27.83	27.70	27.37	26.74
Returns of deposited items	Data not available	6.41	6.38	6.68	6.62	6.45	6.45
Stop payment order	Data not available	25.98	25.70	24.80	25.16	24.89	24.73
Foreign ATM	1.83	1.48	1.37	1.29	1.20	1.17	1.14
ATM surcharge	1.70	1.64	1.61	1.57	1.57	1.58	1.60
Overdraft transfer fee	5.64	4.46	4.28	4.02	4.15	4.38	4.31
ATM annual fee	--	1.22	1.07	0.80	0.83	0.62	0.44
Debit card annual fee	--	0.65	0.58	0.59	0.49	0.94	0.74
Minimum balance to open a non-interest checking account	Data not available	76.43	75.47	63.26	57.76	57.38	55.09
Minimum balance to waive a non-interest checking account fee	749.93	838.85	832.91	796.55	771.00	695.74	670.03

Table 8. Continued

Type of fee	Year 2000	2001	2002	2003	2004	2005	2006
Monthly maintenance fees							
Regular interest checking	12.31	10.75	10.27	9.20	8.66	8.10	7.68
Non-interest checking, balance	8.51	9.73	9.40	9.10	9.03	8.73	8.44
Non-interest checking, flat fee	6.81	6.29	5.48	5.02	5.24	5.49	5.41
Senior checking	13.99	6.99	6.28	5.38	4.87	4.80	4.45
Statement/passbook savings	$5.58	$3.91	$3.80	$3.16	$2.82	$2.80	$2.67

Source: GAO analysis of Informa Research Services data.

Note: Fees are adjusted for inflation and reported in 2006 dollars.

APPENDIX IV. RESOLUTION OF COMPLAINTS RELATED TO FEES AND DISCLOSURES ASSOCIATED WITH CHECKING AND SAVINGS ACCOUNTS

In analyzing the resolution of complaints for fees and disclosures associated with checking and savings accounts, we found similar outcomes among complaints received by the Federal Reserve, FDIC, OCC, and OTS.[1] As shown in figure 10, these federal regulators reported resolving complaints in the following order of decreasing frequency:

1. Finding that the bank was correct. This included instances in which the regulator determined that the financial institution did not err in how it administered its products and/or services to the consumer.
2. Providing the consumer with additional information without any determination of error. This included instances in which the regulator told the consumer that the dispute was better handled by a court or where the regulator determined that rather than wrongdoing there was miscommunication between the bank and its customer.
3. Other, including instances in which the consumer did not provide information needed by the regulator or withdrew the complaint.
4. Determining that the bank was in error. This included instances in which the regulator determined that the bank erred in administering

its products and/or services to the consumer (errors could include violations of regulations).

5. Complaint in litigation, in which the regulator tabled the complaint because it was involved in legal proceedings. This includes instances in which the regulator can not intervene because the issues raised in the complaint are the subject of either past, current, or pending litigation.

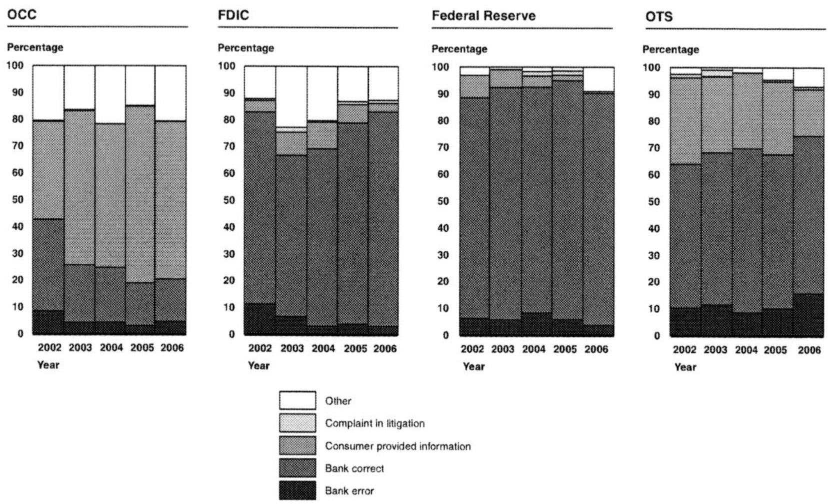

Sources: GAO analysis of OCC, FDIC, Federal Reserve, and OTS data.

Figure 10. Complaint Resolutions Made by Federal Regulators.

REFERENCES

[1] Checking accounts at credit unions are called share draft accounts. For purposes of this report, the use of the term "checking accounts" includes share draft accounts.

[2] 12 C.F.R. § 230.4(b)(4) and Pub. L. No. 102-242, title II, subtitle F, 105 Stat. 2334 (Dec. 19, 1991), *codified at* 12 U.S.C. §§ 4301-4313.

[3] 12 C.F.R. Part 205 and Pub. L. No. 90-321, title IX, as added Pub. L. No. 95-630, title XX, § 2001, 92 Stat. 3728 (Nov. 10, 1978), *codified at* 15 U.S.C. §§ 1693, 1693a-1693r.

[4] The Federal Reserve has responsibility for state-chartered banks that are members of the Federal Reserve System, while the Federal Deposit Insurance Corporation (FDIC) oversees state-chartered banks with federally insured deposits that are not members of the Federal Reserve System. National banks are overseen by the Department of the Treasury Office of the Comptroller of the Currency (OCC), while its Office of Thrift Supervision (OTS) oversees federally chartered and state-chartered savings associations with federally insured deposits. The National Credit Union Administration (NCUA) oversees federally chartered and state-chartered credit unions whose member accounts are federally insured. State-chartered banks, thrifts, and credit unions are also subject to supervision by the state in which they are chartered. This report uses the term "federal banking regulators" to refer collectively to the Federal Reserve, FDIC, NCUA, OCC, and OTS.

[5] 70 *Fed. Reg.* 9127 (Feb. 24, 2005) (OCC, Federal Reserve, FDIC, and NCUA); 70 *Fed. Reg.* 8428 (Feb. 18, 2005) (OTS). We refer to the

joint guidance and OTS guidance collectively as "interagency guidance."

[6] The six states are California, Connecticut, Illinois, Maine, Massachusetts, and New York. We selected these states to illustrate a variety of regulatory efforts and for geographical dispersion.

[7] We reviewed five examinations from each regulator that were selected for dispersion by asset size of the institution and by geography. These examinations, however, are not representative of all federal bank regulators' examinations.

[8] GAO employees followed a standard script and process. If the first or second bank employee encountered did not provide the requested information, or if the GAO employee was instructed to wait, and a period of 10 minutes or more elapsed without the information being provided, we characterized the result of the visit as "unable to obtain the information."

[9] U.S. PIRG, *Big Banks, Bigger Fees 2001, PIRG National Bank Fee Survey* (Washington, D.C.: November 2001).

[10] See GAO, *Financial Regulation: Industry Changes Prompt Need to Reconsider U.S. Regulatory Structure,* GAO-05-61 (Washington, D.C.: Oct. 6, 2004) and GAO, *Financial Regulation: Industry Trends Continue to Challenge the Federal Regulatory Structure*, GAO-08-32 (Washington, D.C.: Oct. 12, 2007) for additional information on oversight of the U.S. financial services industry.

[11] A bank holding company is a corporation that owns or controls one or more U.S. banks. A financial holding company is a bank holding company engaged in a broad range of financial activities, including for example insurance underwriting, securities dealing and underwriting, financial and investment advisory services, merchant banking, issuing or selling securitized interests in bank-eligible assets, or generally engaged in any non-banking activity authorized by the Bank Holding Company Act. *See* 12 U.S.C. § 1841.

[12] 70 *Fed. Reg*. 29582 (May 24, 2005).

[13] *See* 12 C.F.R. Part 707.

[14] Some fees have increased and decreased since 2000, but have an overall increase in the time period analyzed.

[15] GAO analyzed data from two private vendors, Moebs $ervices, Inc. and Informa Research Services, Inc. Moebs $ervices provided data gathered through telephone surveys for each of the years 2000 through 2007, based on statistically representative samples of institutions.

Informa Research Services provided data for each of the years 2000 to 2006. The Informa Research Services data were typically gathered from retail banks with large market shares in specific areas and are not statistically generalizable to other institutions. Because the data provided by Moebs $ervices cover more years and are statistically representative of all depository institutions, we relied on those data primarily to characterize overall trends in fees. For more detailed information on the characteristics of data sets and the data reported by each vendor, see appendixes I and III.

[16] We also obtained this data from Informa Research Services (see app. III). Unless noted otherwise, dollar amounts in the report and figures are shown in 2006 dollars, calculated using the Consumer Price Index calendar year values.

[17] We analyzed data for banks and thrifts in one institution type category because we were unable to obtain data from both Moebs $ervices and Informa Research Services that disaggregated these two institution types.

[18] Board of Governors of the Federal Reserve System, *Annual Report to the Congress on Retail Fees and Services of Depository Institutions* (Washington, D.C.: June 2003).

[19] FDIC-insured institutions are required by statute to report financial data quarterly, known as "Reports of Condition and Income" or "call reports" for banks and Thrift Financial Reports for thrifts, to each institution's primary supervisory agency. These reports provide details on income and certain financial condition information.

[20] Federally insured credit unions are required to report financial information similar to that required for banks and thrifts to NCUA on a quarterly basis. However, credit unions are not required to report on SCDA but are required to report on fee income.

[21] The Federal Reserve's 2007 study of non-cash payments released on December 10, 2007, revealed that in 2006, more than two-thirds of all U.S. non-cash payments were made electronically. From 2003 to 2006, the period covered by the study, all types of electronic payments grew while check payments decreased. The rapid growth in debit card use resulted in the transaction volume of debit cards surpassing that of credit cards for the first time between 2005 and 2006.

[22] Check 21 Act, Pub. L. No. 108-100, 117 Stat. 1177 (Oct. 28, 2003) *codified at* 12 U.S.C. §§ 5001–5018. The act authorizes the use of a new negotiable instrument called a substitute check to facilitate the

broader use of electronic check processing. A substitute check is a paper reproduction of an original check that contains an image of the front and back of the original check, is suitable for automated processing in the same manner as the original check, and meets other technical requirements. See Board of Governors of the Federal Reserve System, *Report to the Congress on the Check Clearing for the 21st Century Act* (Washington, D.C.: April 2007).

[23] According to Federal Reserve officials, the hold periods are designed to cover (1) the time it takes to send the check from the depository institution to the paying institution, (2) the time permitted for the paying institution to determine whether to pay the check, and (3) the time it takes to return an unpaid check from the paying institution to the depository institution.

[24] We conducted an extensive literature review and identified only one study that analyzed consumer's debit card use and its impact on overdraft fees. See Halperin, Eric, et al. *Debit Card Danger: Banks Offer Little Warning and Few Choices as Customers Pay a High Price for Debit Card Overdrafts*, Center for Responsible Lending (Jan. 25, 2007).

[25] 70 *Fed. Reg.* 29582 (May 24, 2005).

[26] Fusaro, Marc, *Consumers Checking Account Behavior: Are "Bounced Check Loans" Really Loans?* Theory, Evidence, and Policy (2007).

[27] Lisa James and Peter Smith, *Overdraft Loans: Survey Finds Growing Problems for Consumers*, Center for Responsible Lending (April 2006).

[28] While the examination procedures for assessing compliance with Regulations DD and E were similar among the five federal bank regulators, the ways in which the regulators conducted compliance examinations varied. Both NCUA and OTS conduct compliance procedures along with safety and soundness procedures during the same examination. The other regulators conduct compliance examinations separately from safety and soundness examinations.

[29] 12 C.F.R. § 7.4002(b) provides considerations that national banks should take into account when setting fees, including (1) establishing fees on a competitive basis and not on the basis of any agreement, arrangement, undertaking, understanding, or discussion with other banks or their officers and (2) establishing the amount of non-interest charges and fees based on business decisions that are made according to sound banking judgment and safety and soundness principles, which

include consideration of the cost of providing the service, the deterrence or misuse of the service by customers, the enhancement of the bank's competitive position, and the maintenance of the safety and soundness of the bank.

[30] See 70 *Fed. Reg*. 9127 (Feb. 24, 2005) for the guidance issued jointly by OCC, the Federal Reserve, FDIC, and NCUA and 70 *Fed. Reg.* 8428 (Feb. 18, 2005) for the guidance issued by OTS. OTS issued separate guidance that, by and large, was similar to the guidance issued by the other federal regulators. Federal credit unions were already subject to certain regulatory requirements governing the establishment and maintenance of overdraft programs under 12 C.F.R. § 701.21(c)(3). This regulation requires a federal credit union offering an overdraft program to adopt a written policy specifying the dollar amount of overdrafts that the credit union will honor (per member and overall); the time limits for a member to either deposit funds or obtain a loan to cover an overdraft; and the amount of the fee and interest rate, if any, that the credit union will charge for honoring overdrafts. The 2005 interagency guidance supplemented but did not change these regulatory requirements for federal credit unions.

[31] NCUA, *Overdraft Protection (Bounce Protection) Programs*, Letter No: 05-CU-03 (February 2005).

[32] Consumers may initially contact a federal regulator about their complaints using various methods, such as telephone, mail, fax, or e-mail. Regulators normally do not formally accept a complaint until they have received written or electronic confirmation of the complaint because many complaints involve personal information about the consumer that the regulator cannot request from a bank without the consumer's consent. All of the federal regulators reported that they had systems in place to refer or forward complaints to the correct regulator if a consumer was unsure of which agency to contact.

[33] See GAO, *OCC Consumer Assistance: Process is Similar to That of Other Regulators but Could be Improved by Enhanced Outreach*, GAO-06-293 (Washington, D.C.: Feb. 23, 2006) for more information on the complaint process used by OCC and the other federal banking regulators.

[34] Generally, the federal regulators can take formal enforcement actions against financial institutions when compliance examiners find violations of laws, rules, or regulations, unsafe or unsound practices, breaches of fiduciary duty, and violations of final orders. Formal

enforcement actions include cease and desist orders, written agreements, removal and prohibition orders, and orders assessing civil money penalties.

[35] While we actually visited 202 branches of depository institutions, we were unable to meet with a depository institution representative in 17 branches, thus, we excluded these branches in our analysis.

[36] For more information on the methodology for our direct observations, see appendix I.

[37] The sample of 185 bank branch visits was not a random, statistically representative sample from the overall population of banks. Therefore, the results we obtained, and the differences in results between subgroups of our sample, are not representative of any larger population of bank branches. See appendix I for information on the design of the sampling and data collection methods for the visits.

[38] U.S. PIRG, *Big Banks, Bigger Fees 2001, PIRG National Bank Fee Survey* (Washington, D.C.: November 2001).

APPENDIX I

[1] Board of Governors of the Federal Reserve System, *The 2004 Federal Reserve Payments Study: Analysis on Non-cash Payments Trends in the United States: 2000–2003* (Washington, D.C.: Dec. 15, 2004). Board of Governors of the Federal Reserve System, *The 2007 Federal Reserve Payments Study: Non-cash Payments Trends in the United States: 2003–2006* (Washington, D.C.: Dec. 10, 2007).

[2] *EFT Data Book: The Complete Guide to the ATM, Prepaid and POS Debit Markets in the United States and Europe*, ATM&Debit News, 2008 Edition (Sept. 27, 2007).

[3] Halperin, Eric, Lisa James and Peter Smith, *Debit Card Danger: Banks offer little warning and few choices as customers pay a high price for debit card overdrafts*, Center for Responsible Lending (Jan. 25, 2007).

[4] Fusaro, Marc, *Consumers Checking Account Behavior: Are "Bounced Check Loans" Really Loans?* Theory, Evidence, and Policy (2007).

[5] James, Lisa and Peter Smith, *Overdraft Loans: Survey Finds Growing Problem for Consumers*, Center for Responsible Lending (April 2006).

[6] Board of Governors of the Federal Reserve System, *Report to the Congress on the Disclosure of Point-of-Sale Debit Fees* (Washington, D.C.: November 2004).

APPENDIX II

[1] We spoke with a merchant trade association, card associations including VISA and MasterCard, card-issuing depository institutions (e.g., banks), and third-party processors supporting merchant acquirers or card-issuing institutions.

[2] Real-time account balance information for the purpose of this report is the most current balance that the depository institution has arrived at when the consumer is conducting their transaction with a debit card at a point-of-sale terminal or ATM. Some depository institutions update their balances throughout the day and call this the available daily balance. Other institutions update the balance once a day (usually in the evening) and post this balance throughout the day.

[3] ATM owners can be banks, merchants, or independent service organizations—companies that specialize in offering ATMs. For more information related to ATM transaction processing and disclosure issues, see GAO, *Automated Teller Machines: Issues Related to Real-time Fee Disclosure*, GAO/GGD/AIMD-00-224 (Washington, D.C.: July 11, 2000).

[4] For additional information on fee and disclosure issues associated with point-of-sale debit transactions, see Board of Governors of the Federal Reserve System, *Report to the Congress on the Disclosure of Point-of-Sale Debit Fees* (Washington, D.C.: November 2004).

[5] To complete a PIN-based debit transaction, a consumer must have a debit card linked to a deposit account at a depository institution that is a member of an EFT network, and the merchant must have a network-compatible point-of-sale terminal.

[6] EFT networks are the telecommunications and payments infrastructure linking consumers, merchants, and banks. The physical components consist of point-of-sale terminals, telecommunications connections, apparatus that route transaction information to appropriate parties, and computers that store deposit and transaction information. According to the 2008 EFT Data Book, as of March 2007, the top 10 EFT networks based on total volume of transactions were Interlink, Star, Accel/Exchange, Interac, Pulse, NYCE, Coop, Jeanie, Shazam, and MoneyMaker.

[7] Historically, the term "acquirer" referred to the depository institution that connected the merchant to the network. Currently, the term refers

either to the merchant's processor—a third-party entity that performs a variety of merchant-related payment activities—or to the depository institution that sponsors the processor's access to the EFT network.

[8] According to Federal Reserve officials, under the overdraft protection programs banks may establish a cushion for paying overdrafts, and they might not pay an overdrawn item if the consumer has exceeded their cushion. In addition, under these programs, banks state that the payment of overdrafts is discretionary. If a consumer has insufficient funds at the time of authorization, but has established overdraft protection by linking their checking account to a savings account, credit card, or line of credit, then the purchase would be authorized, and funds would be transferred from the linked account, with an overdraft transfer fee applied to the transaction.

[9] According to Federal Reserve officials, under card association rules, a consumer will not always be required to sign the receipt. For example, in certain product markets transactions below a defined dollar threshold, which is currently $25, do not always require the customer's signature.

[10] The merchant typically receives payment within 2 days of the transaction. This 2-day range applies to a transaction submitted electronically; a merchant may face a longer delay for the crediting of a paper-based signature debit transaction. Most signature-based debit card transactions are submitted electronically.

[11] According to Federal Reserve officials, an ATM switch will be used to route the transaction if the EFT network used for the transaction is different from the EFT network in which the ATM operator participates.

[12] EFT Data Book, 2008 edition.

[13] The ACH Network is a nationwide batch-oriented electronic funds transfer system that provides for the interbank clearing of electronic payments for participating depository financial institutions. The Federal Reserve and Electronic Payments Network act as ACH operators, central clearing facilities through which financial institutions transmit or receive ACH entries.

[14] However, this option would not help consumers of smaller institutions that do not update the available balance throughout the day.

APPENDIX III

[1] The data from Moebs $ervices and Informa Research Services showed similar trends using nominal dollars, but once we adjusted for inflation, some of the data trend patterns differed between the two data sources. These differences appeared to be due to differences in the characteristics of the institutions surveyed. Informa Research Services had a higher concentration of large institutions in its earlier survey group. Since larger institutions tend to charge higher fees, this would likely result in a higher average fee than if the survey group had a higher proportion of smaller institutions.

[2] In the tables that follow these instances are identified by "- -."

[3] *Interest checking–regular:* A basic interest checking account. *Non-interest checking–balance:* A checking account that does not earn interest and requires a minimum balance in order to waive the monthly maintenance fee. *Non-interest checking–flat fee:* A checking account that does not earn interest and has a monthly fee that cannot be waived with a minimum balance. *Senior checking:* A checking account exclusively for seniors. It can be interest or non-interest bearing. *Statement savings/passbook savings:* Passbook Savings accounts record information in a passbook. A "Statement Savings" account receives a statement in the mail from the bank.

APPENDIX IV

[1] NCUA could not report the outcomes of complaints in the same categories reported by the other regulators.

INDEX

A

B

C

D

E

F

G

H

I

J

L

M

S

T

U

V

W

Y